Special Publication No. 88

# Fine Chemicals for the Electronics Industry II: Chemical Applications for the 1990s

The Proceedings of a Symposium organised by the Applied Solid State Chemistry and the Fine Chemicals and Medicinals Groups of the Royal Society of Chemistry

University of York, 18th-20th April 1990

Edited by

**D. J. Ando**
Queen Mary and Westfield College, University of London

and

**M. G. Pellatt**
BDH Limited, Poole

**British Library Cataloguing in Publication Data**
International Symposium on Fine Chemicals for the
Electronics Industry (2nd 1990 York)
 Fine chemicals for the electronics industry II:
 Chemical applications for the 1990s.
 I. Ando, D. J.  II. Pellat, Dr. M. G.
 621.381

ISBN 0-85186-887-8

# ANDERSONIAN LIBRARY

## 2 7. MAR 91

### UNIVERSITY OF STRATHCLYDE

© The Royal Society of Chemistry 1991

*All Rights Reserved*
*No part of this book may be reproduced or transmitted in any form*
*or by any means—graphic, electronic, including photocopying, recording,*
*taping or information storage and retrieval systems—without written*
*permission from The Royal Society of Chemistry*

Published by The Royal Society of Chemistry
Thomas Graham House, Science Park, Cambridge CB4 4WF

Printed in Great Britain by J. W. Arrowsmith Ltd., Bristol

D
621.38102'8
FIN

# Preface

This volume contains the majority of the invited papers presented at the 2nd International Symposium "Fine Chemicals for the Electronics Industry", which took place at the University of York from 18-20 April 1990. This meeting followed on from the first, highly successful meeting which was held at the University of Bath in 1986, the proceedings of which were published by the Royal Society of Chemistry as Special Publication No. 60. While the previous Symposium reviewed the state of the art for fine chemicals in the electronics industry in the mid 1980s, the York Symposium looked forward to the mid 1990s. Indeed, the organisers adopted a working sub-title of "Chemical Applications for the 1990s" in their planning and all speakers were asked to keep this in mind when putting together their contributions to the meeting and to the book. In addition to emphasing this strong chemical bias, the speakers were asked to give their predictions with regard to specific market demands in the 1990s, and a swift perusal of the papers will indicate that in the main this has been achieved.

This Symposium, which was again jointly organised by the Applied Solid State Chemistry Group and the Fine Chemicals and Medicinals Group, both Subject Groups of the Royal Society of Chemistry, contained an extremely broad range of topics, all presented by international experts in their field. In this present work, the strong interdisciplinary nature of these newer developments is again very apparent, indicating the success of recent specifically funded pre-competitive research schemes such as JOERS in the UK, and ESPRIT and BRITE-EURAM in Europe, in encouraging chemists, physicists, electronic engineers and other specialists to collaborate in scientific projects.

The chapters in the book cover established and well-known materials and processes such as silicon and other semiconductor materials, MOCVD, etc., through to the important fields of liquid crystal and other display techniques. Attention is also given to newer methods for hard copy output, adhesives, optical data storage materials and the rare earths. Papers are included on nonlinear optical and electro-optical materials, reflecting growing interest in this area, in particular for polymers and organic compounds. The final chapter covers the newer and topical, but albeit rather more speculative, field of molecular electronic materials.

In addition to the invited papers contained in this book, some 20 further contributed papers were given at the meeting, to be published in a future edition of "Chemtronics". These were presented at a lively and well-subscribed poster session, which was noteworthy for the extremely high standard of the contributions. Perhaps this had something to do with the award of a cash prize for the best paper! In the event, it proved very difficult for the judges to choose a final winner. Eventually, the prize was awarded to D. H. Zheng and D.C. Bradley of QMW for their paper "Complexes of $PH_3$ with Triarylboron Compounds".

It is also worth commenting here on the relatively high proportion of international delegates attending the meeting, a strong sign of worldwide interest in this field. At the same time the point must be made that the UK electronics industry was very poorly represented at the event. It is hoped, therefore, that this book may assist in stimulating wider UK interest in this important field.

Finally, the editors wish to express their gratitude to the co-members of the Organising Committee, namely Dr P. Bamfield (ICI Colours and Fine Chemicals) and Mr L.P. Vergnano (Consultant), without whose help and enthusiasm this Symposium would not have been such a success. Special thanks also are due to Mrs Elaine Wellingham, of Conference Secretariat, who acted as Conference Secretary and Treasurer and "tied the whole thing together". We also wish to thank the authors for providing their manuscripts, in the majority of cases without too much chasing, and Miss Catherine Lyall (Royal Society of Chemistry)for compiling everything for the purpose of these proceedings.

David J. Ando (Queen Mary & Westfield College)

Martin G Pellatt (BDH Limited)

August 1990

# Contents

## Materials for the Semiconductor Industry

Chemical Processing in the Fabrication of Advanced
Silicon Semiconductor Devices and Structures    3
*A.M. Hodge, A.G. Morpeth, and P.C. Stevens*

Electronic Applications of Indium and Arsenic    36
*R.G.L. Barnes*

Materials for MOCVD and Epitaxial Processes    49
*D.C. Bradley*

## Rare Earth Materials

Rare Earths in the Electronics Industry    63
*K.A. Gschneidner, Jr.*

## Flat Panel Displays and Electronic Printing

High Information Content Display Trends in the 1990s    97
*F. Funada*

The Chemistry of Displays for the 1990s    114
*I.C. Sage*

The Physics of Displays for the 1990s    130
*E.P. Raynes*

Dye Diffusion Thermal Transfer Printing (D2T2)    147
*R.A. Hann*

## Materials for Insulation and Optical Data Storage

New Developments on Isocyanate-based Casting Resins and
Polyurethane Compounds 163
   J. Franke and H.P. Müller

Materials for Optical Data Storage 183
   M. Emmelius, G. Pawlowski, and H.W. Vollmann

## Liquid Crystal Polymers and Nonlinear Optical Materials

Liquid Crystal Polymers 203
   G.W. Gray

Polymers for Nonlinear and Electro-optic
Applications 223
   R.N. DeMartino

Organic Materials for Nonlinear Optical
Applications 237
   D. Broussoux, P. Le Barny, J.P. Pocholle,
   and P. Robin

## Molecular Electonics - 21st Century Fine Chemicals?

Molecular Electronic Materials 265
   D. Bloor

Subject Index 279

**Materials for the Semiconductor Industry**

Chemical Processing in the Fabrication of Advanced Silicon
Semiconductor Devices and Structures
A.M. Hodge, A.G. Morpeth and P.C. Stevens (RSRE)


Electronic Applications of Indium and Arsenic
R.G.L. Barnes (Johnson Matthey plc)


Materials for MOCVD and Epitaxial Processes
D.C. Bradley (Queen Mary Westfield College)

# Chemical Processing in the Fabrication of Advanced Silicon Semiconductor Devices and Structures

A. M. Hodge, A. G. Morpeth,* and P. C. Stevens

ROYAL SIGNALS AND RADAR ESTABLISHMENT, MALVERN, WORCESTERSHIRE WR14 3PS, UK

## 1 INTRODUCTION

The ubiquitous integrated chip, IC, the principle component of modern digital electronics, is found throughout Industry, Commerce, and in the home. In most cases the major component of IC's is the group IV semiconductor element silicon. The silicon microelectronics industry makes a major contribution to the economies of the USA, the EC, and Japan with sales worth $210 billion in 1986[1] and a projected value of $1000 billion by the year 2000.

The expansion of the microelectronics industry is dependent on continuous technological advance epitomised by the rate of device miniaturisation and the exponential growth in the number of components per chip, Figure 1[2,3]. Silicon IC's have developed from the integration of a few transistors, resistors, and other devices in 1958 to complex circuits containing $10^5$–$10^6$ components in the late 1980's. The basic structure of the main element of ICs', the transistor, has changed little. The transistor, Figure 2, is essentially an on/off switch representing one and zero in binary logic. When the gate is biased by an applied voltage, the electric field at the silicon substrate surface under the gate inverts allowing the current to flow, the on state. Changes in the applied voltage at the gate change the transistor state.

As can be seen from Figure 2, the transistor structure consists of several layers which must be produced using a range of processing techniques including: chemical cleaning, controlled oxide growth,

Crown copyright
© Controller, Her Majesty's Stationery Office, London 1990

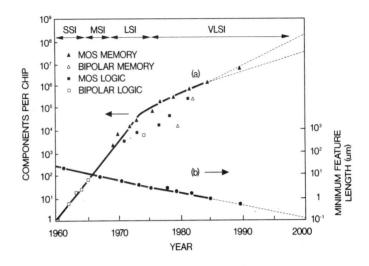

**Figure 1:** The exponential Growth, (a) in the number of components per IC chip and The Exponential Reduction, (b) in device dimensions, *after Moore[2] and Bezold[3]*.

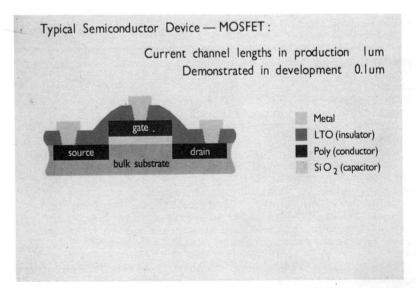

**Figure 2:** Schematic diagram of a simple MOS Transistor.

deposition of thin layers, pattern definition in these layers, and the refinement of electrical characteristics by impurity doping. All the methods highlighted require some chemical input, Table 1 lists example process steps with chemical requirements. From Table 1 it is apparent that within silicon device processing there is a need for numerous gases, solvents, and aqueous media as well as a thorough understanding of the chemistry of such materials.

The purpose of this paper is to present a broad description of chemical processing in silicon semiconductor fabrication. It is not intended that it should be exhaustive and the reader is directed to a number of excellent reviews [1,4-6] for a more detailed discussion of the topics covered in this work.

## 2 GENERAL CHEMICAL REQUIREMENTS

Table 1 illustrated some of the common chemical processing which is a vital part of the fabrication of silicon devices. As devices have become smaller and complexity has increased, Fig. 1, the quality of the chemicals used has become increasingly important. When considering chemicals for the semiconductor industry not only are impurities important but also particulate levels. In a typical fabrication facility, clean room, the atmosphere is tightly controlled, eg temperature 22°C ± 2°C, humidity 45% ± 5%, whilst the air would generally be Class 100, less than 100 particles > 0.5µm diameter per cubic foot of air, and Class 10 - 1 in the operating areas over the equipment. In comparison the highest grade reagent would have very low impurity levels but particle counts of $10^3$ and above; such particulate levels can have a catastrophic effect on the performance and yield of modern silicon ICs.

The demand for high purity low particulate reagents has led to the development of a range of electronic grade chemicals such as Selectipur™, Puranal™, and Isoclean™. Table 2 lists parts of a typical specification for hydrofluoric acid 1% and Acetone [7]. Almost 40 elements, salts, and solvents are specified to parts per million levels or better whilst particulate contamination is controlled to less than 250 particles > 0.5µm per millilitre of reagent. Such specifications are significantly improved when bottled chemicals are replaced by large cylinders for automatic dispensing systems which allow point of use filtering. The gases used in silicon microelectronics are also controlled to

## TYPICAL SERIES OF PROCESS STEPS LEADING TO THE FABRICATION OF A MOSFET

| PROCESS | CHEMICAL REQUIREMENTS |
|---|---|
| Clean | A wet chemical clean at temperature using: $H_2O_2$, $NH_4OH$, $H_2O$; then $H_2O_2$, HCl, $H_2O$. |
| $SiO_2$ Growth | High temperature oxidation in either $O_2$ or $H_2O/O_2$. |
| Polysilicon Gate | Pyrolysis: $SiH_4 \rightarrow Si + 2H_2$ |
| Polysilicon Doping | Diffusion of $POCl_3$ or Implantation |
| Photolithography | Photosensitive polymers, Naphthoquinone Diazides, Quarternary amines |
| Plasma Etching | Low temperature plasma, high temperature electrons, based on chlorine species |
| Implantation | Beams of single mass ions |
| Clean | Post implant as above |
| Inter-layer Dielectric | Gas phase reaction $SiH_4$, $PH_3$, $O_2$: Phosphosilicate glasses |
| Densification ILD | High temperature anneal: doped glass flow |
| Photolithography | See above |
| Plasma Etching | Based on fluorine plasma |
| Interconnect/Contact | Metallisation using sputtering |
| Photolithography | See above |
| Plasma Etching | Based on $BCl_3$, $Cl_2$ plasma |
| Intermediate Steps | Photoresist stripping: several routes eg fuming $HNO_3$, proprietary resist removers, $O_2$ plasma<br>Various wet etches eg HF dips; HF, $HNO_3$, $H_2O$ mixes, acid etching of metal |

Table 1

### ACETONE
### VLSI Selectipur®

LSI Particle Class I
$C_3H_6O$

Certificate of Guarantee

| | |
|---|---|
| Assay (GC) | min 99.5% |
| Resistivity | min 5 M$\Omega$·cm |
| Arsenic (As) | max 0.01 ppm |
| Boron (B) | max 0.01 ppm |
| Cadmium (Cd) | max 0.01 ppm |
| Iron (Fe) | max 0.05 ppm |
| Gold (Au) | max 0.02 ppm |
| Potassium (K) | max 0.02 ppm |
| Copper (Cu) | max 0.01 ppm |
| Sodium (Na) | max 0.1 ppm |

### HYDROFLUORIC ACID 1%
### VLSI Selectipur®

LSI Particle Class I
HF

Certificate of Guarantee

| | |
|---|---|
| Assay (acidimetric) | 1.0 ± 0.05% |
| Hexafluorosilicate ($SiF_6$) | max 50 ppm |
| Sulphate ($SO_4$) | max 1 ppm |
| Arsenic & Antimony (as As) | max 0.02 ppm |
| Boron (B) | max 0.02 ppm |
| Cadmium (Cd) | max 0.01 ppm |
| Iron (Fe) | max 0.1 ppm |
| Gold (Au) | max 0.02 ppm |
| Potassium (K) | max 0.05 ppm |
| Copper (Cu) | max 0.01 ppm |

Upwards of 25 other elements, salts and organics are specified.

Table 2

a very high purity. Gases such as argon, silane, phosphine are required at the 99.9995% purity level with in line filtering to 0.2µm.

The general requirements for chemical processing in silicon semiconductor fabrication therefore cover: solvents, aqueous media, and gases, all of which must be manufactured to very high purity levels with controlled particulate contamination. Having highlighted the type, purity, and quantification of typical reagents used in silicon microelectronics it is now important to discuss in detail some of the chemistry of processing. To do this the remainder of the paper will be split into three main topics covering: the chemistry of pattern transfer, safe replacements of toxic chemicals, and current issues and future requirements.

## 3 THE CHEMISTRY OF PATTERN TRANSFER

Figure 2 illustrated a simple transistor; to achieve the structure shown requires a minimum of 15 processing steps the majority of which involve layer deposition followed by pattern definition and pattern transfer. These three areas, deposition, lithography, and etching provide the major components of pattern transfer and each technique has its own associated chemistries.

### Deposition

The principal techniques used in silicon processing for the deposition of high purity thin films are: sputtering, a physical process used mostly for metal films, and the chemical vapour deposition, CVD, of silicon, silicon dioxide, silicon nitride, and some metals. The former of these techniques, sputtering, requires little chemical input being based on the bombardment of a target by argon species. The bombardment leads to the transfer of kinetic energy from the argon atom/ion to a target atom which is then ejected onto a substrate. Conversely CVD as the name suggests is very dependent on the chemistry of gas species and consequently much of this discussion will be concerned with this technique.

The four main methods of chemical vapour deposition, Table 3, are atmospheric CVD, low pressure CVD, plasma-enhanced CVD and photo-enhanced CVD; of these the second, LPCVD, is most often found in silicon microfabrication. The latter two techniques are more usually research topics, although in recent times the

## THE PRINCIPLE DEPOSITION TECHNIQUES USED IN DEVICE FABRICATION

| Deposition Technique | Acronym | Temperature |
|---|---|---|
| Atmospheric Chemical Vapour Deposition | APCVD | 900-1200C |
| Low Pressure Chemical Vapour Deposition | LPCVD | 400-850C |
| Plasma Enhanced Chemical Vapour Deposition | PECVD | 200-350C |
| Photo Enhanced Chemical Vapour Deposition | Photo-CVD | 200-350C |

## TYPICAL DEPOSITION REACTIONS

| Product | Reactants | Temperature C | Product | Reactants | Temperature C |
|---|---|---|---|---|---|
| $SiO_2$ | $SiH_4 + CO_2 + H_2$ | 850-950 | $Si_3N_4$ | $SiH_4 + NH_3$ | 700-900 |
| | $SiCl_2H_2 + N_2O$ | 850-900 | | $SiCl_2H_2 + NH_3$ | 650-750 |
| | $SiH_4 + N_2O$ | 750-850 | Plasma $Si_3N_4$ | $SiH_4 + NH_3$ | 200-350 |
| | $SiH_4 + NO$ | 650-750 | | $SiH_4 + N_2$ | 200-350 |
| | $Si(OC_2H_5)_4$ | 650-750 | Plasma $SiO_2$ | $SiH_4 + N_2O$ | 200-350 |
| | $SiH_4 + O_2$ | 400-450 | Silicon | $SiH_4$ | 570-670 |

Table 3

need to reduce thermal budgets, the degree of high temperature processing a silicon device receives, and increased complexity have led to their introduction in silicon foundaries. Atmospheric CVD has in most applications been superseded by LPCVD.

Table 3 also lists a number of archetypal CVD processes covering a range of temperatures and products. Common starting materials include silane dichlorosilane, ammonia, and nitrogenous oxides. One organosilicon derivative is also used, specifically for the deposition of silicon dioxide; this is tetraethyl orthosilicate, TEOS. Of the reactions described, Table 3, two will be discussed in detail these are: the deposition of silicon for polysilicon gate materials, Scheme 1a, and the formation of phosphosilicate glasses, Scheme 1b.

$$SiH_4 \longrightarrow Si + 2H_2$$

Scheme 1a

$$SiH_4 + 4PH_3 + 6O_2 \longrightarrow SiO_2 + 2P_2O_5 + 8H_2$$

Scheme 1b

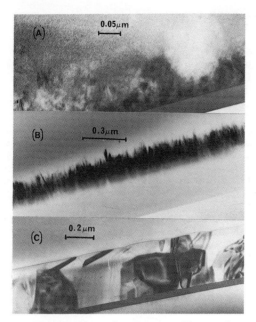

**Figure 3:** Cross-sectional TEM studies of a-Si (a), C-Si, (b) and recrystallised a-Si, recrystallisation at 900°C for 30 mins in $O_2POCl_3$ Ambient (c)

These reactions are major steps in the fabrication of the transistor, Figure 2. Deposited silicon provides the gate material which controls the switching of the device, phosphosilicate glasses isolate the devices and prevent short circuits within the transistor.

## The Pyrolysis of Silane

Silane pyrolysis is the most important method of silicon deposition used in the semiconductor industry. In the low pressure regime the reaction takes place at 570-670°C using either pure silane or silane in a carrier gas which may be hydrogen, nitrogen, argon, or helium. The structure of the deposited layer is highly dependent on the deposition temperature[1, 8-10]; thus at 570°C the silicon is wholly amorphous with no long range order whilst at temperatures above 620°C the deposited layer is polycrystalline showing both a columnar structure and some long range orientation of crystals. The two types of film are usually referred to as α-Si and c-Si, amorphous and polycrystalline silicon respectively; both types of silicon give crystallisation and grain growth on heating, Figure 3[11].

The rate of deposition of silicon is also dependent on the temperature of processing. This temperature dependence is usually exponential and follows the Arrhenius equation:

$$R = A.exp(-qE_a/kT)$$

where $R$ is the deposition rate, $E_a$ is the activation energy in eV, $T$ is the absolute temperature in °K, $A$ is the frequency factor, $k$ the Boltzmann constant, and $q$ is the electronic charge. If the logarithm of deposition rate is plotted against the inverse of temperature a straight line results, slope $-qE_a/k$, Figure 4[1]. In most cases the deposition rate increases with temperature; however at high dilutions of silane the reaction becomes mass transport limited and deposition rate falls away, cf. Fig. 4 deposition rate at silane partial pressure of 20 and 10mτ, as rate of reaction exceeds rate of surface absorption.

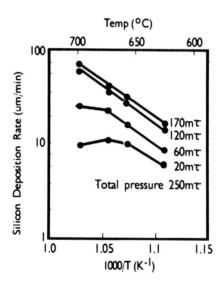

**Figure 4:** Arrhenius plot for polysilicon deposition at different silane partial pressures.

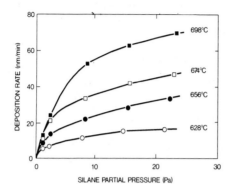

**Figure 5:** The effect of silane concentration on polysilicon deposition rate at different temperatures.

Other variables which can affect the deposition process include pressure, silane concentration, and dopant concentration. If total gas flow, the sum of the partial pressure of silane, carrier gas, and dopant if present, is varied then deposition rate becomes a linear function of pressure; however, if only the flow of carrier gas is changed then rate depends only slightly on pressure. The effect of silane concentration on deposition rate is non-linear, Figure 5[1]. This behaviour is believed to be the result of the absorption processes occuring on the growing surface, Scheme 2, Figure 6[12, 13].

$$SiH_{4(g)} \rightarrow SiH_{2(g)} + H_2$$

$$SiH_{2(g)} \rightarrow SiH_{2(ad)}$$

$$SiH_{2(ad)} \rightarrow SiH + H_{2(ad)}$$

$$H_{2(ad)} \rightarrow H_{2(g)}$$

Scheme 2

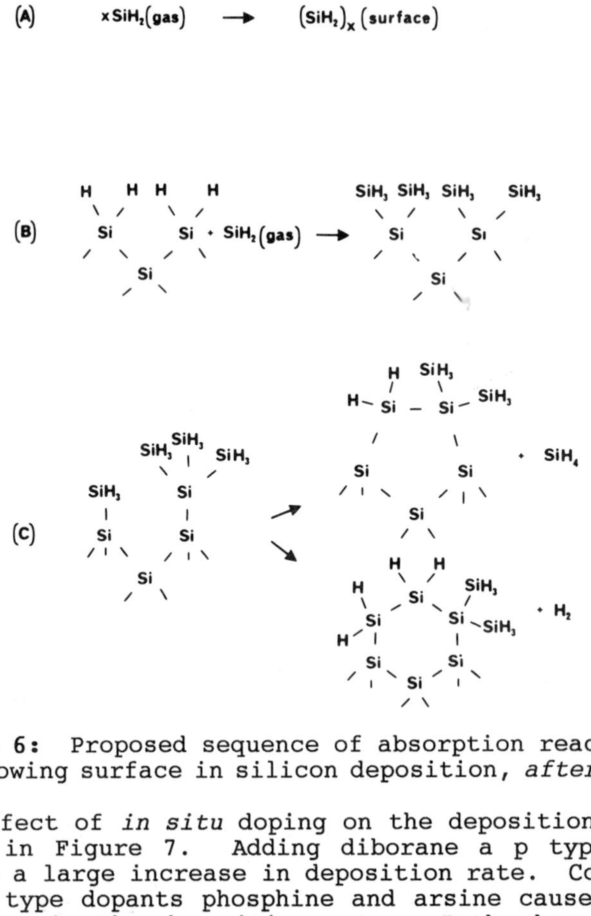

(A) $x\,SiH_2(gas) \longrightarrow (SiH_2)_x\,(surface)$

(B) [diagram of surface Si-H species + $SiH_2$(gas) → extended surface with $SiH_3$ groups]

(C) [diagram showing further reaction producing $SiH_4$ and $H_2$]

**Figure 6:** Proposed sequence of absorption reactions at the growing surface in silicon deposition, *after Scott*[12].

The effect of *in situ* doping on the deposition rate is shown in Figure 7. Adding diborane a p type dopant causes a large increase in deposition rate. Conversely the n type dopants phosphine and arsine cause a rapid decrease in the deposition rate. Both phosphine and arsine are strongly absorbed onto the growing surface thus blocking the absorption of silane as silene, Figure 6, and so reducing the deposition rate. Diborane however forms boron radicals, $\cdot BH_3$, which could catalyse the gas phase reactions in Scheme 2 and so increase the deposition rate.

The preceding paragraphs have served to demonstrate the chemical complexity of an apparently simple process, the conversion of silane into silicon by deposition techniques. Another important deposition process is the formation of phosphosilicate glasses and some discussion of this reaction now follows.

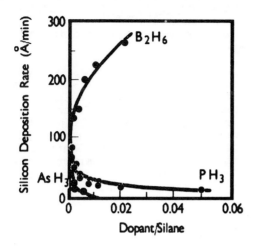

**Figure 7:** The effects of dopants on the polysilicon deposition rate at 610°C.

## The Low Temperature Gas Phase Reaction of Silane, Phosphine, and Oxygen

Deposited silicon dioxide films are used in microelectronics as: insulation, passivation, implantation masks and diffusion masks or when doped: diffusion sources and flow glasses for topography smoothing.

There are several methods of silicon dioxide deposition at varying temperatures, Table 3. These include the reaction of silane and oxygen at 400 - 450°C, the pyrolysis of tetraethyl orthosilicate TEOS, at 650 - 750°C, and the reaction of dichlorosilane with nitrous oxide at 900°C. Each method has advantages and disadvantages. Thus the use of silane and oxygen at low temperatures allows oxide to be deposited over metal layers but does not provide conformal step coverage, whilst the deposition of silicon dioxide from TEOS gives a uniform layer regardless of topography. Lastly the high temperature process using dichlorosilane and nitrous oxide while providing excellent uniformity does lead to chlorine incorporation which can cause a deterioration of the underlying layers. Of the methods described the chosen route, despite the step coverage problems, is usually deposition from silane/oxygen mixtures. The main reason for this is the low temperature of deposition which allows oxide to be put down over metal layers.

When the low temperature oxide, LTO, is used as the interlayer dielectric between deposited silicon and the first level metal the poor step coverage is overcome by doping. LTO is doped by introducing phosphine into the gas mixture; this produces a phosphosilicate glass, P-Glass, which when annealed at high temperature in either an oxygen atmosphere or an inert ambient will flow, Figure 8 [14]. The importance of this characteristic lies in the smoothing of topography allowing subsequent layers to be deposited over a surface with less abrupt step changes.

The deposition of LTO is dependent on the same variables as polysilicon; temperature, pressure, reactant concentration, and dopants. It is however a more complex process and other less definitive variables

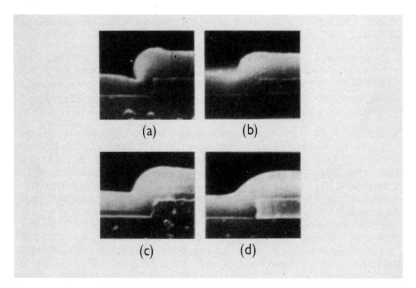

Figure 8: SEM cross-section, mag. 10K, of P-glass samples annealed in steam at 1100°C for 20 mins, weight %P: (a) 0.0, (b) 2.2, (c) 4.6, and (d) 7.6, *after Adams and Capio*[14].

such as wafer spacing, total and localised gas flows; reactant distributions and gas phase transport also influence the control of the deposition process.

The complexity of the process may be highlighted by considering the dependence of film formation on oxygen concentration, Figure 9[15]. At constant temperature if oxygen concentration is increased the deposition rate increases rapidly, passes through a maximum then gradually falls. Surface reactions cause this behaviour; at high oxygen concentrations the surface is almost saturated with oxygen and consequently further silane reactions are blocked. Similar effects are observed when phosphine is introduced as a dopant; in this case the rate rapidly decreases then slowly increases. This behaviour is also attributed to surface absorption effects; phosphine is strongly absorbed blocking surface active sites and so slowing further reaction.

When the properties of deposited silicon dioxide are considered the layer is found to be amorphous consisting of $SiO_4$ tetrahedra[16]. Bound within the silicon-oxygen network is hydrogen, this is most prevalent in $SiO_2$ formed at low temperatures, 350 - 450°C, and is present as silanol, SiH, and water[17]; typically LTO contains 1-4% silanol and less than 0.5% SiH. The density of the film is variable, 2.0 - 2.2g/cm$^3$, and heating of the deposited film causes densification, decreasing the oxide thickness, increasing the density and ordering the $SiO_4$ tetrahedra. The oxide will also react with atmospheric moisture. This is particularly important in phosphosilicate glasses where high concentrations of phosphorus absorb moisture to form phosphorus based acids which corrode subsequent layers.

So far this section has described some of the chemical vapour deposition routes to thin films in the fabrication of silicon devices. However, these films are not selective and cover the whole of the substrate surface. The formation of devices requires the selective removal of the deposited film from much of the substrate and in the next two parts of this section the chemistry of this process will be described.

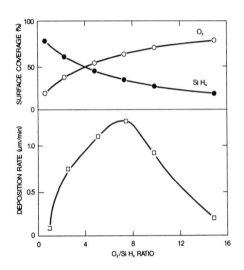

**Figure 9:** The deposition rate and surface coverage for the silane-oxygen reaction at 350°C and atmospheric pressure, *after Cobianu and Pavelescu*[15].

Lithography:   The Art of Pattern Definition

Microcircuit fabrication requires that the components of the circuit are fashioned as a series of structures in a range of thin layers. The overall device is therefore built up as successive layers are deposited and patterned. The first stage in the patterning is lithography in which the desired pattern is generated in a resist layer. The resist is usually a polymeric film spin-coated onto the substrate and containing radiation sensitive groups which chemically respond to incident radiation, eg light, electrons, X-ray or ions. Exposure of the resist forms a latent image of the circuit pattern which is subsequently developed by either wet or dry processing producing a 3-dimensional relief image.

The various stages of the lithographic process are shown schematically in Figure 10. In this example a photosensitive resist is exposed when *uv* light passes through a mask made up of clear and dark features which define the circuit. Those areas of the resist which are illuminated become either soluble, positive resist, or

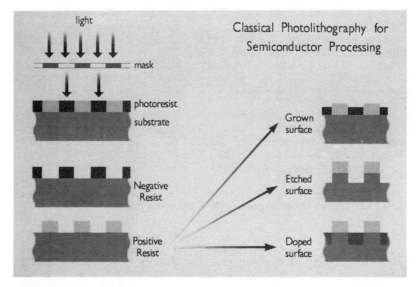

**Figure 10:** Schematic representation of the stages of the photolithography process.

insoluble, negative resist, in a specific developer. After the resist has been developed the underlying layer may be patterned or doped in the exposed regions whilst the resist protects the remaining area of the substrate.

The role of a lithographic resist is therefore two-fold: first it must react to the exposing radiation forming a latent pattern of the required circuitry, secondly the resist regions remaining after development must withstand subsequent processing. To perform both these functions the resist requires a range of properties viz:

i) Easy application with good adhesion
ii) Uniform & reproducible thickness
iii) Sensitive to the exposing radiation with high resolution
iv) Rapid development process
v) Resistant to the pattern transfer process
vi) Readily removed <u>only</u> after pattern transfer

To satisfy all the criteria described resist technology has developed a range of chemistries[18] covering simple one-component materials through to multi-component systems which include polymers, photo-active molecules and complex amines.

Figure 11[4] highlights the photochemical reactions of two example resists, one positive and the other negative. In the case of positive resist an aqueous, base soluble, novolac binding polymer contains a 1,2-naphthaquinone diazide substituted in the benzene ring with a bulky hydrophobic group as a dissolution inhibitor. On exposure to *uv* light the diazide breaks down to form a ketene via a keto-carbene which in the

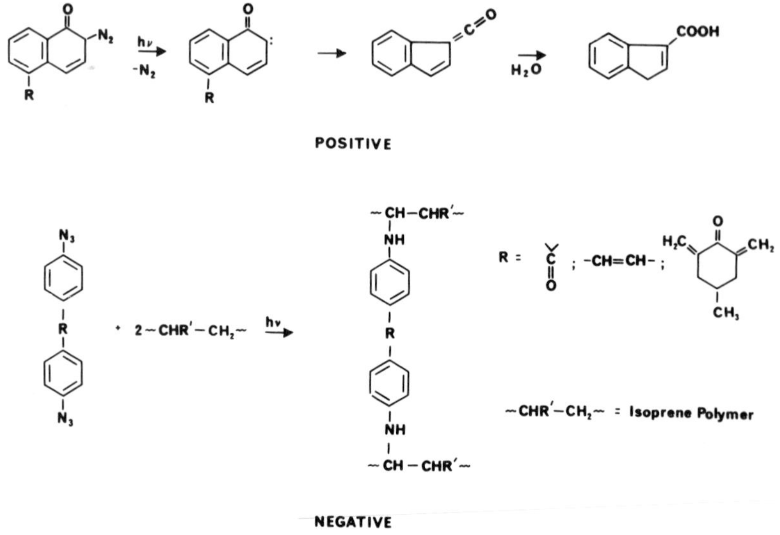

**Figure 11:** Principal photochemical reaction of a positive and a negative photoresist, *after Ledwith*[4].

presence of water forms the indene carboxylic acid. The acid no longer prevents dissolution allowing exposed region of resist to dissolve in the developer. A typical positive resist formulation will contain up to 40% by weight diazide.

In negative resists where irradiated areas become insoluble a common system involves aromatic azides and cyclised isoprene polymers. Aromatic azides readily decompose on exposure to yield highly reactive nitrene intermediates which undergo insertion reactions with aliphatic carbon-hydrogen bonds. Consequently when irradiated in the presence of polyisoprene the bis-azido compound produces crosslinks which give insoluble polymer matrices, Figure 11[4]. These matrices then form the latent image whilst the unexposed regions are dissolved.

Lithography therefore provides a further example of the dependence of silicon microfabrication on detailed knowledge of important chemistries. In particular the process of pattern definition is controlled by the reactions of irradiated molecules, and the behaviour of polymeric mixtures when processed in the harsh environment of the next stage of pattern transfer, etching.

Plasma Etching

The final stage of the pattern transfer process is etching. In etching the substrate exposed by lithography is removed; by a chemical reaction, physically, or quite often by a combination of these methods. Furthermore, the substrate should be removed in such a way as to minimise the loss of underlying layers and with minimal damage to both the lower layer and the resist. Consequently etching must be highly selective in the majority of applications and this requires several different chemistries.

The two principal techniques used in pattern transfer are wet, isotropic, and dry, plasma, etching. Figure 12 illustrates these two processes. In isotropic etching there is no preferred direction and the substrate is etched equally in all directions, whilst anistropic etching is highly directional and produces vertical features. The differing results from wet and dry etching lead to the use of aqueous methods for coarse geometries, >10µm and bulk substrate removal only; readers are referred to the work of Kern[4,19,20] and

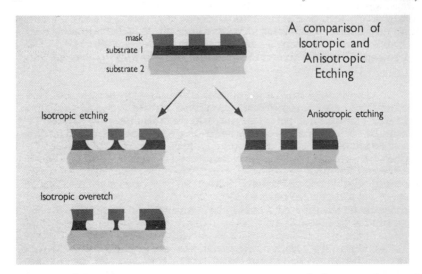

**Figure 12:** Schematic representation of isotropic and anistropic etching and the results of an isotropic overetch.

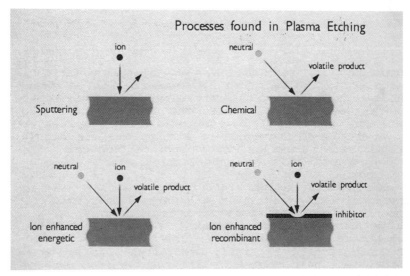

**Figure 13:** The four basic mechanisms of plasma etching, after Flamm[4].

Pliskin [17] for detailed treatment of these techniques. The remainder of this section will be concerned with plasma etching which has in most cases superseded wet etching.

A simple plasma etching reactor consists of opposed parallel plate electrodes in a low pressure, 0.1 - 1τ, chamber. When a high frequency is applied between the electrodes current flows and a plasma forms. Mounting semiconductor substrates on the electrode surface exposes them to a range of reactive neutral and charged species from the plasma. Some of the plasma species either react with the substrate forming volatile products which are pumped away, or simply eject substrate atoms by transfer of energy, sputtering. The overall result is that those areas of substrate exposed by the lithographic process, see above, are eroded thus transferring the resist pattern to the substrate.

Though a complex process the basic mechanism of plasma etching may be broken down into four categories[4] viz:

i) Sputtering
ii) Chemical etching
iii) Energy driven ion-enhanced etching
iv) Inhibitor protected sidewall ion-enhanced etching

Each of the four categories, Figure 13, show different characteristics any combination of which could be present in a plasma etching process. In sputtering impinging particles, usually positive ions strike the substrate with high kinetic energy. Some of that energy is transferred to a substrate atom which is then ejected from the surface. There is a nett removal of the substrate in what is a purely mechanical process.

Chemical etching requires an active species from the gas phase to react at the surface giving a **volatile product**. Product volatility is essential to prevent the build up of a protective layer which would inhibit further reaction and hence slow or stop further etching. Chemical etching is, as would be expected, isotropic, cf. Fig.12, and highly selective. It is often used in the closing stages of an etch when the interface between layers is approached.

In the latter two categories fast directional etch rates are achieved when there are fluxes of both neutral and ion species present at the substrate surface. The effect is to give removal rates which are greatly in excess of the sum of separate chemical and sputter etching. In the energy driven mechanism little or no etching occurs when only neutral species are present. Ion bombardment causes damage, eg formation of dangling bonds, lattice disruption, reactant injection, bond cleavage, which increases the reactivity of the substrate surface allowing neutral species to effect material removal.

In the inhibitor-protected sidewall mechanism the role of impinging ions is to remove a "protective" inhibitor film from the surface of the substrate. The exposed area of material then reacts with incoming neutral species giving a volatile product.

The mechanisms occurring during any specific plasma etch are dependent not only on the chosen chemistry, an example of which will be given later, but also on a range of equipment specific parameters. The reason that plasma control must be based on equipment parameters relates almost exclusively to the non-equilibrium nature of a plasma discharge. The plasma is a dynamic system in which active species are not well defined and subject to constant variations in concentration[21]. The five main discharge parameters used to control plasma etching and the aspects of the process they affect are listed viz:

| | | |
|---|---|---|
| i) | Pressure | Ion & Electron Energies; Kinetics; Transport |
| ii) | Power | Ion & Radial Densities; Etch Rate |
| iii) | Excitation Frequency | Electric Field; Plasma Fluctuations; Ion Energies |
| iv) | Temperature | Etch Rate; Selectivity; Anisotropy |
| v) | Flow Rate | Residence Times; Diffusion Rates; Chemical Reaction Rates |

One of the most well understood processes in plasma etching is the isotropic etching of silicon by atomic fluorine, Figure 14. This process is used to illustrate how a single active species in a plasma may generate two or more differing reactions which lead to the erosion of a substrate surface.

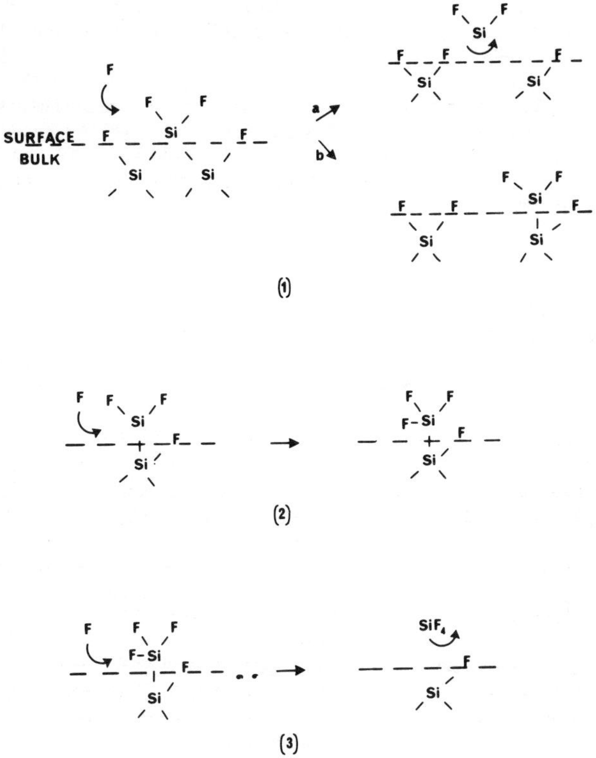

Figure 14: Basic reactions found in the isotropic etching of silicon by atomic fluorine, *after Flamm*[4].

In the process a passivating $SiF_2$ surface must first be penetrated by an incident F atom to precipitate attack at subsurface Si-Si bonds. Once the Si-Si bonds are cleaved the etch proceeds in one of two ways. Volatile gaseous $SiF_2$ may be generated directly, 1a, or the fluorine may be chemisorbed giving bound -$SiF_2$-, 1b, which is fluorinated further to produce $SiF_3$ and $SiF_4$, 2 and 3. It is apparent therefore that the mechanism of the reaction is a two channel concerted reaction resulting in the direct formation of gaseous $SiF_2$ and a bound fluorosilicon radical. The kinetics show that 1a and 1b are branches of a single rate limiting step possibly involving a common intermediate [4,22-24].

In the example of atomic fluorine etching a silicon substrate it has been shown that the reactions of a plasma are not always well defined and a single intermediate may be the source of two or more rate determining steps. In silicon processing however plasma etching is not usually limited to only one active component; more often two or more gas sources are fed into the reactor during the same stage of etching. Examples of these more complex systems include $O_2$ and $CHF_3$; Ar and $BCl_3$; HCl, $BCl_3$ and $Cl_2$; $CF_4$, $BCl_3$, $Cl_2$, and HCl.

Detailed explanation of the chemistries and process requirements which have led to such complex mixtures are beyond the scope of this paper. However the importance of chemistry in identifying how the various components of any one etch system interact with one another and with the substrate cannot be overstated. Nor can the difficulties encountered in transferring an etch process from one equipment to a second dissimilar equipment be understated.

Throughout the preceding section an attempt has been made to highlight the intimate relationship between chemistry and the fabrication of a typical silicon semiconductor device. By using the chemistry of pattern transfer as an example a range of different chemical reactions have been shown to play a major role in the day to day processing of a silicon foundry. These include apparently simple gas phase interactions through to photochemistry and the behaviour of highly reactive non-equilibrium species. However, the chemistry of silicon device fabrication is not without its disadvantages. In particular many of the critical processing steps require the use of highly toxic, corrosive, flammable, or merely harmful chemicals. In the next section some of the new developments leading to the replacement of certain dangerous chemicals will be described.

3 Replacement of Toxic Chemicals in Semiconductor Processing

Table 4 lists many of the chemicals used for the

fabrication of silicon semiconductor devices; most are either toxic or flammable whilst some are also either pyrophoric or corrosive. Recently a gradual trend has developed seeking to replace some of the more dangerous chemicals with a range of "safer" alternatives. In this section a description of some of these alternatives will be given.

## PROPERTIES OF GASES COMMONLY USED IN SILICON SEMICONDUCTOR FABRICATION

| Gas | Formula | Hazard | Flammable Limits | Exposure Limit (ppm) |
|---|---|---|---|---|
| Ammonia | $NH_3$ | Toxic, Corrosive | 16-25 (%vol Air) | 25 |
| Argon | Ar | Inert | - | - |
| Arsine | $AsH_3$ | Toxic | - | 0.05 |
| Diborane | $B_2H_6$ | Toxic, Flammable | 1-98 (%vol Air) | 0.1 |
| Dichlorosilane | $SiCl_2H_2$ | Flammable, Toxic | 4-99 | 5 |
| Hydrogen | $H_2$ | Flammable | 4-74 | - |
| Hydrogen Chloride | HCl | Corrosive, Toxic | - | 5 |
| Nitrogen | $N_2$ | Inert | - | - |
| Nitrogen Oxide | $N_2O$ | Oxidant | - | - |
| Oxygen | $O_2$ | Oxidant | - | - |
| Phosphine | $PH_3$ | Toxic, Flammable | Pyrophoric | 0.3 |
| Silane | $SiH_4$ | Flammable, Toxic | Pyrophoric | 0.5 |

Table 4

Deposition Processes

In the case of deposition processing attempts have been made to replace the very hazardous gaseous sources such as silane, arsine, phosphine, with "safer" liquid sources sited within the furnace system. The attraction of such sources are not purely related to the reduction in hazard, though in some cases it is quite significant, but more typically because homogeneous surface reactions, particularly in silicon dioxide deposition, can be obtained. Consequently the proposed alternatives give improved conformality and enhanced step coverage when compared with the more traditional processes. The improvement however cannot be achieved without an increase in the degree of higher temperature processing a device must withstand.

The other benefits of using liquid sources relate to the considerable reduction in the complexity of the storage and delivery systems necessary to manage safely processes currently relying on gases such as silane, diborane, and phosphine. These gases often require elaborate purpose designed gas storage compounds. Such compounds require: flow limiting devices, double wall electro-polished continuously purged stainless steel pipework, advanced gas monitoring devices, and in the case of fire or a release of gas fail safe isolation. All are features which occupy a considerable amount of valuable facility "real estate". They are also costly installations to run and maintain, and can significantly add to the nuisance value to the local neighbourhood especially extract fan noise, regular tanker deliveries, and heightened concern about risks.

The liquid sources are either based on a simple "bubbler" concept in which the vapour is transported into the furnace system often using nitrogen as the carrier gas, or a vacuum induced vapour draw method. They take up little room as they can usually be located in the furnace source cabinets attached to the main furnace unit. There are however drawbacks: the chemicals in use, Table 5, are clearly organic, and can be the source of carbon contamination which could lead to device deficiencies and yield reductions.

One process which has been shown to accept liquid replacements for both silane and phosphine with limited contamination risk is the deposition of silicon dioxide. The "safer" method is accomplished by substituting tri-*ter*-butyl phosphine, TBP, or trimethylphosphite, TMPI, for phosphine and tetramethyl-cyclotetrasiloxane, TOMCATS™, for silane.

The TOMCATS $SiO_2$ deposition process is based on the following chemical reaction:

$$C_4H_{16}Si_4O_4 + 10 O_2 \xrightarrow{550-600°C} 4 SiO_2 + 8 H_2O + 4 CO_2$$

This method gives films which it is reported have much lower stresses than layers deposited using tetraethyl orthosilicate TEOS and are deposited at a lower temperature, 550-600°C for TOMCATS; 700-750°C for TEOS, thus reducing the impact on the process thermal budget. However in terms of temperature alone neither process can compete with the silane LTO process which requires temperatures of around 400°C to form the deposition products.

## Semiconductor Processing: Alternative "Safer" Source Materials for Device Fabrication

| PROCESS | GASES USED | "SAFER" REPLACEMENT |
|---|---|---|
| LPCVD POLYSILICON | SILANE | NONE - OXYGEN WOULD BE REQUIRED TO SCAVENGE THE CARBON FROM CURRENTLY AVAILABLE SOURCES |
| DOPED "POLY" | AS ABOVE + PHOSPHINE | TMPI - TRIMETHYLPHOSPHITE |
| LPCVD LOW TEMP. OXIDE | SILANE | TOMCATS - TETRAMETHYLCYCLOTETRA-SILOXANE |
|  | OXYGEN | NONE |
| PROCESS DOPANTS: | DIBORANE | TMB - TRIMETHYLBORATE |
|  | PHOSPHINE | TMPI - TRIMETHYLPHOSPHITE |
| LPCVD OXIDE | SILANE | TEOS - TETRAETHYL-ORTHOSILICATE TOMCATS - TETRAMETHYLCYCLOTETRA-SILOXANE |
|  | OXYGEN | NONE |
| ATMOSPHERIC OXIDATION | OXYGEN | NONE |
|  | HYDROGEN | NONE |
|  | HYDROGEN-CHLORIDE | TCA - TRICHLOROETHANE |

Table 5

Phosphine source substitution, noted above, is achieved by using either TBP or TMPI as the dopant. However, although TBP is considerably less toxic than phosphine, it is a registered pyrophoric substance. In effect therefore other considerations must be taken into account, ie storage, handling, locality of other sources in use which may be released in the case of fire before it can be said that a safer substitute for phosphine has been installed into the process matrix.

Another common dopant used in LTO reflow processes is diborane which is highly toxic and pyrophoric. A suitable substitute can be found in trimethylborate which is considerably less toxic and is not pyrophoric.

Lastly, to date no satisfactory alternative to silane has been identified for use in the deposition of silicon gate materials.

## Lithography Processes

The outcome of a recent occupational hygiene study[25] has prompted a lot of general concern amongst workers in lithography areas who are continually exposed to trace amounts of glycol ethers. The major area of concern specifically relates to the employment of female staff of child bearing age, who by the very nature of the lithography process will be exposed to low levels of materials which have been shown to exhibit a teratogenic nature when tested on animals.

The study in question did not contain sufficient employees to be statistically significant, and contained data of doubtful accuracy. However it has prompted a large multi-million dollar study in the USA into the welfare of employees in the semiconductor industry, which should make its final report during 1991.

The particular compound causing concern is 2-ethoxyethyl acetate, commonly called 2-EEA or Cellosolve acetate™. 2-EEA is used as the solvent in a range of propriety photoresist formulations for optical lithography processes. An alternative to 2-EEA is propylene glycol monomethylether acetate, which is currently incorporated into only one of the safer photoresists on the market; those others on which there is information still contain 2-EEA.

## Aqueous and Gaseous Acids

The final area in which there is a demand for less toxic alternatives involves the range of acids used in the semiconductor processing. Very few silicon devices are fabricated without the use of solutions containing hydrochloric, hydrofluoric, nitric or sulphuric acids, nor can patterning occur without the use of gases such as hydrogen chloride, boron trichloride, hydrogen bromide, or fluorine. Hydrogen chloride is also used in silicon oxidation usually as a cleaning agent and occasionally during oxide growth.

To date only hydrogen chloride has proved

replaceable and then only in development studies. The substitute used was trichloroethane which replaced HCl in high temperature oxidations of silicon. The major problem with this material is the high probability of carbon contamination and the carcinogenicity of the 1,1,2-isomer.

Attempts have also been made to replace hydrochloric and hydrofluoric acid using the halogenated derivatives of acetic acid. Unfortunately all such derivitives are themselves highly corrosive, though not as destructive as hydrofluoric acid, and have the added disadvantage of providing a source of hydrocarbon films.

In closing this section it is worth noting that the semiconductor industry is somewhat conservative in its approach, with large proving runs being required by most device houses in order to qualify the product before its introduction. Therefore if no direct process advantage is offered by a replacement chemical for any of the extremely hazardous chemicals used in the industry then bulk take up will be small without the back up of substantial, well founded safety data, or the introduction of legislation covering its use.

## 4  CURRENT ISSUES AND FUTURE TRENDS

In this section of the paper a personal view of some of the present needs and future directions of chemical processing in the semiconductor industry will be given. The three main areas where chemistry can make an impact are: alternative "safer" source materials, novel sources for new techniques and higher purity chemicals.

Table 6 lists examples of replacement sources for silane, hydrogen chloride, diborane, and phosphine which have been used in standard CVD processes. All these alternatives have associated hazards which could preclude their use in a silicon foundary. Consequently a large body of chemical research on safer CVD sources could still be undertaken as could the adaption of known compounds to new processes. The principal requirements of such materials are: cost, low toxicity levels, low flammability combined with high volatility, minimised contamination especially carbon, and stability.

Stability is very important as liquid source could be in place for as much as twelve months and throughout that time it must retain its composition. Any

## HAZARDS ASSOCIATED WITH ALTERNATIVE SOURCES

| "SAFER" ALTERNATIVE | HAZARD |
|---|---|
| Tetraethyl orthosilicate $((EtO)_4Si)$ | Flammable, Toxic, Readily Oxidised |
| Trichloroethane $(C_2H_3Cl_3)$ | Toxic, Possible Carcinogen |
| Trimethylborate $((CH_3O)_3B)$ | Flammable, reacts vigorously with Water/Steam |
| Trimethylphosphite $((CH_3O)_3P)$ | Flammable |

### Table 6

deterioration in the purity of any one source can have catastrophic effects on the integrity of processing. Equally important is the combination of low flammability and high volatility; these two characteristics dictate both processing and storage. Any source which is not easily delivered to the process equipment by current methods would be seen as uneconomic, whilst flammable materials require specialist storage and isolation.

An alternative to safer current processing is the use of novel techniques with new sources, in particular plasma and photo-enhanced CVD, two methods which are established in research applications and emerging as foundry processes. Present PECVD and Photo-ECVD routes are based on LPCVD reactions, Table 3. Hence they are non-ideal and there are significant problems with eg hydrogen incorporation, particulate contamination, variations in film characteristics[4] all of which are dependent on equipment parameters. Consequently there is a need for new sources which allow greater control of processing and reduced variation across batches of wafers. The ideal source for PECVD or Photo-ECVD should therefore be safe with a well defined reaction route, stable over a range of conditions, show limited secondary reactivity i.e. generate only very few active species, and give volatile waste products.

Another area in which chemistry is making an important contribution is the development of metal CVD[26], in particular LPCVD of aluminium and a range of refractory metals. Current metallisation techniques, sputtering and evaporation, are line of sight methods and this leads to quite severe step coverage problems.

LPCVD metallisation is conformal giving a uniform thickness regardless of topography.

The most widely studied process is the deposition of tungsten via hydrogen and silicon reduction of tungsten hexafluoride[27-30] and many good results have been obtained. However, several problems remain particularly when the deposition is intended to be selective, i.e. over silicon windows in silicon dioxide layers. Among the problems encountered are: poor adherence, non-uniform depositions, encroachment, and tunnelling. Many of the difficulties highlighted stem from the reaction sequence, Scheme 3, which produces gaseous $SiF_x$ species;

$$WF_6 + 3H_2 \longrightarrow W_{(s)} + 6HF$$

hydrogen reduction

$$2WF_6 + Si_{(s)} \longrightarrow 2W_{(s)} + 3SiF_4$$

silicon reduction

### SCHEME 3

these may attack the silicon substrate or the silicon dioxide leading to loss of layer integrity. Other routes to tungsten CVD include the reduction of tungsten hexachloride or tungsten hexacarbonyl both of which can give satisfactory layers. However, alternative sources may provide better layers and this is an area in which further studies would be of benefit.

Several processes now exist for the deposition of metals other than tungsten including: aluminium from tri-*iso*-butyl aluminium[31,32], molybdenum from $MoF_6$[33], metal silicides from refractory metal halides and silane[26], and cobalt from cobalt carbonyls[34]. The development of these and the tungsten reactions mentioned above provide great scope for detailed chemical investigations which could lead to significant advances in silicon semiconductor technology.

Lastly as the size of silicon devices shrinks even further with a concomitant increase in complexity the basic chemicals of fabrication, gases, solvent and aqueous reagents, must continue to improve in both purity and particulate contamination. Such improvements must be driven by the manufacturers of high quality electronic grade chemicals, with the full support of all users.

## 5 SUMMARY

In the Semiconductor Industry chemistry and applied physics are intimately related.

Layer processing is a chemical problem.

Many support processes in silicon device fabrication are also chemical.

The type, purity, and particulate levels of process chemicals is vital.

Current source materials are hazardous, as are many of the alternatives, hence there is a need for new sources.

Plasma and photo-enhanced chemical vapour deposition are important techniques which require new chemistries.

*The authors wish to thank the organising committee of the 2nd International Symposium on "Fine Chemicals for the Electronics Industry" for their invitation to present our work. We would also like to thank Her Majesty's Government for permission to publish, and the staff of the Royal Signals and Radar Establishment for their assistance in preparing this manuscript.*

*The following figures are reproduced with the permission of the publishers: Figs. 1, 4, 5, 7, 8, 9, and Table 4: McGraw-Hill International. Figs. 6, 11, 12, 13, 14; Blackie & Son.*

## References

1. "VLSI Technology" 2nd Edition, Ed S M Sze, McGraw-Hill International 1988.
2. G Moore, "VLSI What Does the Future Hold", *Electron. Aust.*, 42, 14 (1980).
3. G Bezold and R Olsen, "The Information Millenium: Alternative Futures", Information Industry Association 1986.
4. "The Chemistry of the Semiconductor Industry", eds S J Moss and A Ledwith, Blackie 1987.
5. E A Irene, "Models for the Oxidation of Silicon", *CRC Critical Rev. in Solid State and Mater. Sci.*, 14 (2), 175 1988).
6. "Introduction to Microlithography", eds L F Thompson, C G Wilson and M J Bowden, *A.C.S.Symp. Ser.* 219 (1983).
7. MERCK Electronic Chemicals, VLSI Selectipur™ Process Chemicals Catalogue 1988.
8. G Harbeke, L Krausbauer, E F Stigmeier, A E Widmer, H F Kappert and G Neugebauer, "Growth and Physical Properties of LPCVD Polycrystalline Silicon Films" *J. Electrochem. Soc.* 131 (3), 675 (1984).
9. G Harbeke, L Krausbauer, E F Steigmeier, A E Widmer, H F Kappert and G Neugebauer, "LPCVD Polycristalline Silicon: Growth and Physical Properties of In-situ Phosphorus Doped and Undoped Films", *RCA Rev.* 44, 287 (1983).
10. T I Kamins, "Polycrystalline Silicon for Integrated Circuit Applications", Kluwer Academic Publishers 1988.
11. A G Cullis and A G Morpeth, Unpublished Results.
12. B A Scott in "Semiconductors and Semimetals", ed J I Pankove, Academic Press 1984, pp 123.
13. F J Kampas in "Semiconductors and Semimetals", ed J I Pankove, Academic Press 1984, pp 261.
14. A C Adams and C D Capio, "Planarisation of Phosphorus Doped Silicon Dioxide", *J. Electrochem. Soc.*, 128, 423 (1981).
15. C Cobianu and C Pavelescu, "Silane Oxidation Study: Analysis of Data for $SiO_2$ Films Deposited by Low Temperature Chemical Vapour Deposition", *Thin Solid Films*, 117, 211 (1984).
16. N Nagasima, "Structure Analysis of Silicon Dioxide Films Formed by Oxidation of Silane", *J. Appl. Phys.*, 43, 3378 (1972).

17. W A Pliskin, "Comparison of Properties of Dielectric Films Deposited by Various Methods", *J. Vac. Sci. Technol.*, 14, 1064 (1977).
18. A Reiser, "Photoreactive Polymers: The Science and Technology of Resists", Wiley-Interscience 1989.
19. W Kern, "Wet-chemical Etching of $SiO_2$ and PSG Films, and an Etching Induced Defect in Glass Passivated Integrated Circuits", *RCA Rev.*, 17, 196 (1986).
20. W Kern and C A Deckert in "Thin Film Processes", eds J L Vossen and W Kern, Academic Press 1978.
21. D L Flamm and G K Herb in "Plasma Materials Interactions", 1, eds D M Manos and D L Flamm, Academic Press 1987.
22. T J Chuang, "Infrared Chemiluminescence from $XeF_2$-Silicon-Surface Reactions", *Phys. Rev. Lett.* 42, 815 (1979).
23. T J Chuang, "Electron Spectroscopy Study of Silicon Surfaces Exposed to $XeF_2$ and the Chemisorption of $SiF_4$ on Silicon", *J. Appl. Phys.*, 51, 2614 (1980).
24. M J Vasile and F A Stevie, "Reaction of Atomic Fluorine with Silicon: The Gas Phase Products", *J Appl. Phys.*, 53, 3799 (1982).
25. (a) Semiconductor Industry Association; "Report of the SIA Task Force Review of the (unpublished) Digital Equipment Corporation Study of Semiconductor Manufacturing Workers" Feb. 19th 1987; (b) H Pastides, E J Calabrese, D W Hosmer and D R Harris, "Spontaneous Abortion and General Illness Symptoms among Semiconductor Manufacturers", *J. Occupational Med.*, 30 (7), 543 (1988).
26. A Sherman, "Chemical Vapour Deposition for Microelectronics: Principles, Technology and Applications", Noyes Publishing 1987.
27. M L Green and R A Levy, "Structure of Selective Low Pressure Chemical Vapour Deposited Films of Tungsten", *J. Electrochem. Soc.*, 132(5), 1243 (1985) and references within.
28. E K Broadbent and W T Stacey, "Selective Tungsten Processing by Low Pressure CVD", *Solid State Technol.*, Dec 1985, p 51 and references within.
29. R S Blewer, "Progress in LPCVD Tungsten for Advanced Microelectronics Applications", *Solid State Technol.* Nov. 1986, p 117 and references within.
30. H Itoh, T Moriya, and M Kashiwagi, "Tungsten CVD: Applications to Submicron VLSICs", *Solid State Technol.*, Nov. 1987, p 83 and references within.

31. M J Cooke, R A Heinecke, and R C Stern, "LPCVD of Aluminium and Al-Si Alloys for Semiconductor Applications", *Solid State Technol.*, Dec. 1982, p 62.
32. M L Green, R A Levy, R G Nuzzo and E Coleman, "Aluminium Films Prepared by Metal-organic Low Pressure Chemical Vapour Deposition", *Thin Solid Films*, 114, 362 (1984).
33. N Lifshitz and M L Green, "Comparative Study of the Low Pressure Chemical Vapour Deposition Processes of W and Mo", *J Electrochem. Soc.*, 135, (7), 1832 (1988).
34. Queen Mary & Westfield College, personal communication.

# Electronic Applications of Indium and Arsenic

R. G. L. Barnes

JOHNSON MATTHEY PLC, ORCHARD ROAD, ROYSTON, HERTFORDSHIRE SG8 5HE, UK

1 ABSTRACT

The post-transition elements of the Periodic Table form a group of metals and metalloids of enormous technological importance. This arises from the characteristics of the energy band-gaps of their various alloys, which can be tailored across the entire range of useful values for electronic activity. A general description of this phenomenon is given, with a more detailed account of the stringent conditions which have to be realised in practice using indium and arsenic as specific examples. Purity requirements, analytical demands, and various approaches to component manufacture, are all discussed.

2 BACKGROUND

The elements of the Periodic Table include a number of metals and metalloids which lie to the right of the transition metals. These have enormous technological importance, and are grouped as shown in Figure 1. Their outer electronic configurations span the stable half-filled shell of four electrons found in silicon and germanium. As in carbon, these hybridise to the tetrahedrally directed $sp^3$ configuration, which confers a diamond-like structure and a modest energy gap between the occupied valence band and unoccupied conduction band energy levels. It is the characteristics of this band gap which make these materials semiconductors, with valuable electronic properties.

*Electronic Applications of Indium and Arsenic* 37

Figure 1

## Relevant Section of the Periodic Table

|  | $s^2$ | $s^2p$ | $s^2p^2$ | $s^2p^3$ | $s^2p^4$ |
|---|---|---|---|---|---|
|  |  | Al | Si | P | S |
| $3d^{10}$ | Zn | Ga | Ge | As | Se |
| $4d^{10}$ | Cd | In |  | Sb | Te |
| $5d^{10}$ | Hg |  |  |  |  |

$sp^3$
Tetrahedral
Coordination

 

The same stable electronic configuration occurs in binary compounds of a Group III and a Group V, or a Group II and a Group VI element. However, the bond formed between them has a partial ionic character; this ionicity is greater the smaller and more highly charged the formal ions involved. The higher the ionicity, the larger is the energy band gap, which means that the range of possible compounds can provide a range of electronic properties, illustrated in Figure 2. This versatility is made complete by moving to solid solutions, or "alloys", of different binary compounds, giving ternary or even quaternary compounds. Because excitation of valence electrons across the band gap can be brought about by the absorption of photons of sufficient energy, these materials provide a sensitivity to a wide range of visible and infra-red wavelengths. The reverse process provides corresponding emission. Applications exploiting this range are listed in Table 1.

Figure 2

## Energy Band-Gaps

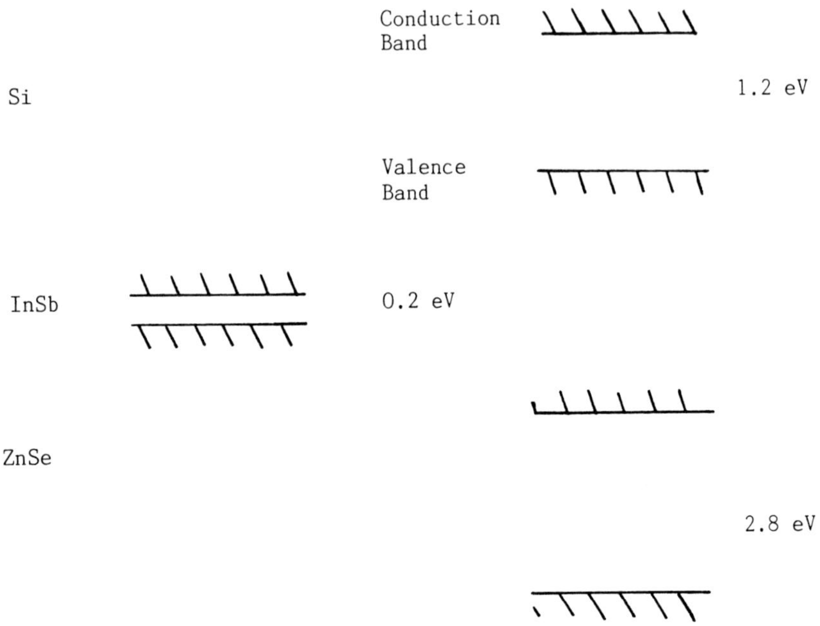

Si — Conduction Band / Valence Band — 1.2 eV

InSb — 0.2 eV

ZnSe — 2.8 eV

Table 1

## Semiconductor Compositions and Applications

| Application | Materials | |
|---|---|---|
| Electronic Components | Si | |
| | Ge | |
| | GaAs | (Ga,Al)As |
| Light-Emitting Diodes | InP | |
| | GaAs | |
| Photo-Detectors | Si | |
| | GaAs | |
| Infra-Red Detectors | InSb | |
| | CdTe | (Cd,Hg)Te |
| Luminescent Components | ZnS | (Zn,Cd)S |
| | ZnSe | |
| Photovoltaics | Si | |
| | CdS | |
| | GaAs | |

## 3 TECHNOLOGICAL NEEDS

A simple energy band gap is in fact an idealised situation, to be found only in totally pure materials with a defect-free crystal structure. In reality, impurities and defects do exist in even the best materials, and these can cause local perturbations in the position of the valence or conduction band edges, or introduce additional energy levels within the band gap. This is shown schematically in Figure 3. The effect of these is to interfere with the simple behaviour described earlier, and to change the optical and electronic properties of a particular material. Sometimes these changes can be beneficial, and then need to be controlled; but accidental misbehaviour can destroy the usefulness of semiconductor materials, and has to be eliminated as far as possible.

Figure 3

Localised Engergy Levels

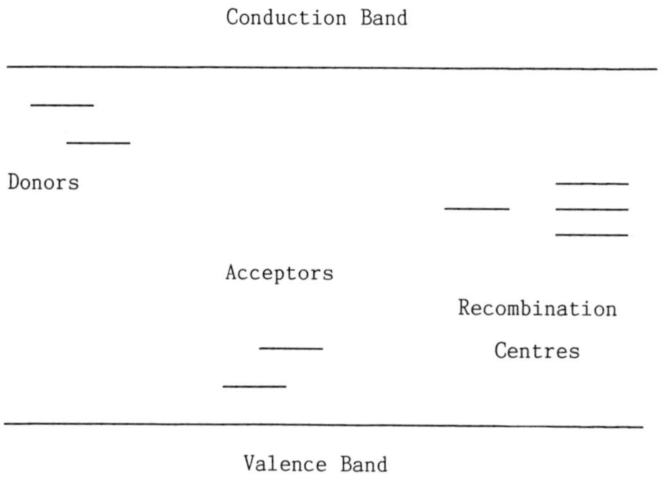

Conduction Band

Donors

Acceptors

Recombination Centres

Valence Band

Band edge perturbations reach extreme proportions at gross concentrations of defects, such as are found at surfaces and grain boundaries. "Band bending" (see Figure 4) completely alters the local characteristics of a material, and is again highly undesirable.

The need to constrain all possible sources of interference with desired behaviour leads to a description of the technological needs of a useful semiconductor material: it has to be of high purity, high structural integrity and with high constitutional uniformity. These conditions are found together in single crystals, and it is in the form of single crystals that compound semiconductors find their applications. With total impurity contents of a few parts per million and very tight stoichiometry tolerances, the degree of control over the chemistry and processing technologies which is required can, perhaps, be appreciated.

Figure 4

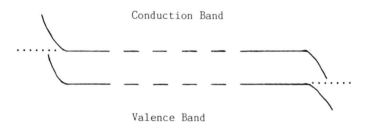

Band Bending

## 4 PURITY REQUIREMENTS

Specific purity requirements can be provided by determining the levels at which semiconductor performance becomes impaired. As an example, tolerable silicon impurity in GaAs is typically less than $10^{15}$ ions per $cm^3$, which is equivalent to approximately 0.05 parts per million by weight. To realise these levels, very careful and sophisticated chemical processing is necessary.

Purification techniques have to recognise and deal with three types of impurity source: raw materials; the natural environment surrounding the chemical operation; and the artificial environment of the process plant, including the materials of which the plant is constructed, and any chemical reagents which might be used en route. Processing techniques themselves can be divided into those which achieve general purification from a number of impurities, and those which are specific for a very few chemically related impurities. The attainable purification levels for the various combinations of impurity source and processing are shown in Table II.

Table II

### Attainable Purification Levels

| Impurity Source | Purification Technique | |
| --- | --- | --- |
| | General | Specific |
| Raw Material | 10 ppm | 0.1 ppm |
| Natural Environment | 0.5 ppm | 0.01 ppm |
| Artificial Environment | 5 ppm | 0.05 ppm |

Protection from recontamination after purification is essential. Figure 5 illustrates the effect of particulate contamination (airborne, for example) on purity. Roughly

Figure 5

### Effect of Extraneous Particulate Matter on Purity

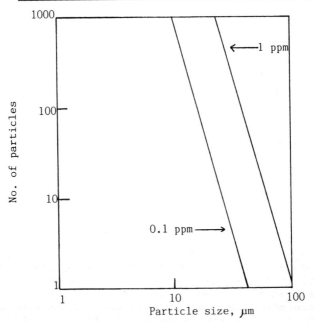

speaking, one 50 micron dirt particle in 1 gram of pure product is equivalent to 0.1 parts per million of impurities. At least a well controlled environment, if not clean room conditions, are necessary, and fully enclosed chemical operations have to be envisaged.

All of this means that very high purity is not cheap. The inter-relationship between quality, performance and cost is shown schematically in Figure 6. The two curves will be in different relative positions for different products, but in general the point of maximum cost-effectiveness lies at their maximum separation. Of course, performance thresholds have to be crossed for viable applications; and better performance at no additional cost is always being sought from materials manufacturers.

Figure 6

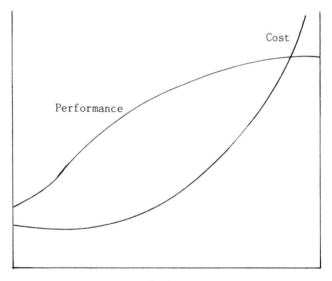

Cost-Effectiveness of Purity

Quality

Knowledge of the purity of these materials demands appropriate analytical techniques. Extreme sensitivity is essential even to detect the very low impurity levels we are concerned with; and freedom from interference from background impurities is necessary for meaningful information. A list of modern analytical methods is given in Table III. Preferred methods are those which require minimal sample preparation, and which can analyse a material directly rather than diluted in solution. Multi-element techniques also have huge cost benefits over single element techniques. Arc emission spectrophotometry and spark source mass spectrometry are well established, with the advantages just described. Glow discharge mass spectrometry is a much more recent advance, offering inherently lower detection limits, but is still enormously expensive.

Table III

### Modern Analytical Methods

**Requiring Sample Preparation and Dilution**

        Absorption Spectrophotometry   (AA)
        Emission Spectrophotometry    (ICP)
        Polarography
        Voltammetry
        Ion Chromatography

**Requiring Long Times with Incomplete Elemental Range**

        Activation Analysis

**Preferred**

        Arc Emission Spectrophotometry
        Spark Source Mass Spectrometry
        Glow Discharge Mass Spectrometry

Typical impurity levels found in 6N (99.9999%) and 7N arsenic and indium (major ingredients of gallium arsenide and indium phosphide respectively) are shown in Table IV. The different impurity "fingerprints" reflect the different chemical behaviour and processing routes for these two examples.

Table IV

Typical Impurity Levels in High Purity
Indium and Arsenic

|      | 6N As | 7N As | 6N In | 7N In |
|------|-------|-------|-------|-------|
| Al   | 0.01  |       | 0.05  | 0.01  |
| Bi   | 0.01  |       | 0.02  |       |
| Ca   | 0.01  |       | 0.04  | 0.02  |
| Cr   |       |       | 0.03  |       |
| Cu   | 0.01  |       | 0.08  | 0.01  |
| Fe   | 0.1   |       | 0.05  | 0.02  |
| Mg   | 0.1   | 0.02  | 0.02  |       |
| Mn   |       |       | 0.02  |       |
| Ni   |       |       | 0.04  |       |
| Pb   |       |       | 0.1   |       |
| Sb   |       |       | 0.03  |       |
| Si   | 0.03  | 0.02  | 0.1   | 0.02  |
| Tl   |       |       | 0.2   |       |

## 5 APPLICATION NEEDS

Sufficiently pure component materials have to be converted to compound semiconductors in single crystal form. Obviously, scrupulous cleanliness still has to be sustained, including in what are effectively fabrication processes. Single crystals can be produced either in bulk form, or in thin layers, by appropriate techniques. However, bulk crystals have to be cut into wafers before use, so that thin layer techniques have major advantages. The best quality is found in bulk crystals, so for the most stringent requirements cut wafers are still needed. The general classes of single crystal growth techniques are given in Table V.

Table V

"Atomic Engineering" Techniques

| Melt Growth | – | Czochralski |
|             |   | Bridgmann |
| Liquid Phase Epitaxy |   | (LPE) |
| Vapour Phase Expitaxy |   | (VPE) |
| Molecular Beam Epitazy |   | (MBE) |

Because control of structure and composition is being achieved down to atomic monolayer proportions, these techniques can be regarded as "atomic engineering" quite literally.

High performance applications of compound semiconductors such as gallium arsenide demand the best electronic properties, and these are found for GaAs in epitaxial layers grown at relatively low temperatures on single crystal substrates. Molecular beam epitaxy (MBE) has emerged as a major technique for epilayer fabrication, utilising effusion of the constituent elements from suitable sources and their directed deposition on the substrate, all under high vacuum conditions. To exploit MBE to the full, more is demanded of an ideal source material than bulk purity alone. Sophisticated control of epilayer growth requires control of the elemental beams; while product reproducibility in a manufacturing operation requires tight control of variables throughout both short-term and long-term operation. This latter requirement means that the local distribution of residual impurities within a source should be uniform, and should also be uniform from one source charge to another over extended periods of time.

The former requirement, that of beam control, requires geometrical control of the effusing surface, without the perturbations which might arise from, for example, a discontinuous source. So a solid, dense source is required, which is normally cylindrical in geometry. To improve beam uniformity during consumption of a source charge, some shaping in the form of a regular slight taper, or succession of different tapers, is an added demand.

The fabrication step is carried out by a vapour phase deposition process in a specially designed quartz vessel. This is chemically polished internally by a proprietary technique, to leave a surface onto which arsenic does not key as it condenses. The condensation zone is engineered to conform quite closely to the cross-section required of the finished charges. Vapour phase transport from the source end of the reactor to the product end is achieved by use of a carefully controlled thermal gradient which can be caused to travel along the reactor as the vapour transport progresses. Once complete, the product is a long rod which shrinks free from the quartz envelope. Its density is typically 99% of theoretical. The entire process takes place in a specially built glove box system. Sub-division of the uniform, high purity rod into MBE

Figure 7a

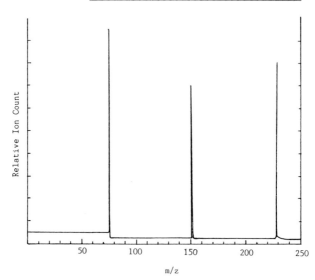

Figure 7b

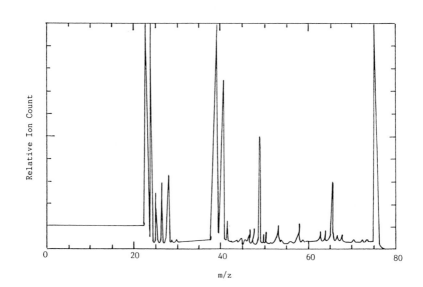

charges with tight dimensional tolerances is accomplished by cutting and trimming using a high power $CO_2$ laser. The cleanliness of this operation is contrasted with mechanical cutting in Figure 7. Verification of suitability for use is obtained by growing GaAs epilayers and measuring their performance. Hall measurement results are summarised in Figure 8.

Figure 8

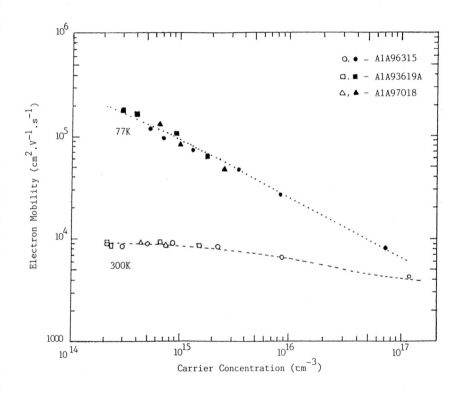

REFERENCES AND ACKNOWLEDGEMENTS

More detailed background may be found in:

"Semiconductor Opto-Electronics" by T.S. Moss, G.J. Barrell and B. Ellis (Butterworths, 1973).

Other information is from work by Johnson Matthey plc over an extended period of time.

LIMA measurements were made by Loughborough Consultants, and Hall data were obtained by R.S.R.E., whose assistance is gratefully acknowledged.

# Materials for MOCVD and Epitaxial Processes

D. C. Bradley

DEPARTMENT OF CHEMISTRY, QUEEN MARY AND WESTFIELD COLLEGE, MILE END ROAD, LONDON E1 4NS, UK

1 INTRODUCTION

Although it is more than twenty years since Hal Manasevit demonstrated that pure gallium arsenide could be deposited by MOCVD using trimethylgallium and arsine,[1] it is only during the past ten years that inorganic chemists have seriously taken up the challenge of synthesizing metallo-organic precursors and physical chemists have addressed the complex problem of unravelling the mechanism of the deposition. Purity is the name of the game and for several years it was asserted by the practitioners of molecular beam epitaxy (MBE) that the "dirty" process of MOCVD could not produce electronic grade materials of comparable quality. The immense strides taken by inorganic chemists in recent years in perfecting the purification has shown that the MOCVD process can match the quality of III-V compound semiconductors deposited by MBE with the added advantage of requiring cheaper equipment and being more amenable to large scale production.

The purpose in designing metallo organic precursor molecules is to enhance the volatility of the element being deposited by attaching suitable organic groups.[2] For example, the boiling point of elemental gallium is >2000°C whereas the boiling point of $GaMe_3$ is only 56°C.

Suitable metallo organic precursors are now available for the deposition of a wide range of refractory materials including metals, metal oxides, metal nitrides, metal sulphides, metal fluorides and a vast range of compound semiconductors.

Since these materials are being used in the high technology industries of electronics, opto-electronics, ceramics, glasses and a range of surface coatings, they must constitute a growing market for fine chemicals.

## 2 COMPOUND SEMICONDUCTORS

Although we are living in the era of the silicon chip there are some specific areas in electronic device manufacture where a compound semiconductor (e.g. GaAs) has distinct advantages especially with regard to electron mobility and optoelectronic properties. Some examples of the uses of III-V, II-VI, IV-VI etc. compound semiconductors are given in Table 1.

Table 1.   Uses of Compound Semiconductors

| | Semiconductors | Uses |
|---|---|---|
| III-V | GaAs/$Al_xGa_{1-x}As$ | Field Effect Transistors, |
| | InP | High Electron Mobility Structures, |
| | $Ga_xIn_{1-x}As$ | Light Emitting Diodes, |
| | $Al_xIn_{1-x}As$ | Photocathodes, Laser Diodes, |
| | $Ga_xIn_{1-x}As_yP_{1-y}$ | Quantum Well Laser Diodes, Integrated Optoelectronics, Wave Guides, Solar Cells |
| II-VI | $Zn\ S_x Se_{1-x}$ | Electroluminescent Displays, |
| | $Cd\ Se_x Te_{1-x}$ | Infrared Detectors and |
| | $Cd_x Hg_{1-x} Te$ | Imaging Devices, Solar Cells, Phosphors, Waveguides |
| IV-VI | PbS, PbSe, PbTe | Infrared Emitting and |
| | SnS, SnSe, SnTe | Detecting Devices. |
| | $Sn_xPb_{1-x} Te$ | |

The preparation and purification of metalloorganic precursors for compound semiconductors has recently been reviewed comprehensively by Jones[3] and will be dealt with fairly briefly here. The electrical and optical properties of III-V semiconductors are sensitive to certain impurities (Si, Zn, C) at concentrations around $10^{14} cm^{-3}$ (e.g. parts per billion) and this posed a problem for the producers of the organometallic precursors containing aluminium, gallium and indium in the form of the trimethyls or triethyls. In addition these compounds are exceptionally air-sensitive posing severe safety hazards and the requirement of special precautions in handling. Furthermore, the Group V element hydrides ($PH_3$ and $AsH_3$), which are extremely toxic gases, were also required to be exceptionally pure because the molecular ratio of $PH_3$ to $InMe_3$ in a typical MOCVD process for depositing InP epitaxially at 650° may be as high as 50:1. Earlier reports of premature reaction of $PH_3$ with $InMe_3$ in the carrier gas ($H_2$) stream with deposition of polymeric materials before reaching the heated substrate and consequent depletion of the indium precursor concentration posed additional problems. The problem was solved by using the volatile adducts $InMe_3$.L (L = $NMe_3$, $NEt_3$, $PMe_3$, $PEt_3$) which could be readily purified and were much less chemically reactive than $InMe_3$.[4] The fact that the adducts for gallium and indium are much closer in volatility (e.g. $GaMe_3$. $NMe_3$, b.p. 164°C, $InMe_3$. $NMe_3$, b.p. 171°C) than the trimethyls ($GaMe_3$, b.p. 56°C, $(In Me_3)_4$, b.p. 136°C) was an added bonus especially for the deposition of a ternary compound such as $Ga_x In_{1-x}As$ where accurate control of the gallium and indium precursor concentrations is essential. Fortunately the tertiary bases $NR_3$ or $PR_3$ are so stable that they take no part in the decomposition of the metal trialkyls and function simply as carriers.

Our greatest success in precursor purification resulted from the use of non-volatile Lewis bases such as $PPh_3$, diphos, triphos, and tetraphos[5] which has led to the commercial exploitation of the diphos adducts.[6] Diphos (1,2-bis-diphenylphosphino ethane) forms crystalline 2 : 1 adducts $(MR_3)_2$ diphos (M = Al, Ga, In; R = Me, Et) which release the pure metal trialkyl on warming in vacuo.

$$2MR_3 + diphos \rightleftharpoons (MR_3)_3 \; diphos \quad - - - (1).$$

The diphos adducts, which are reasonably stable in air and thus convenient for storage and transportation, can be purified to a very high degree by crystallisation from hydrocarbon solvents. Using trimethylindium generated from commercial (Epichem Ltd) diphos adduct Thrush et al.[7] have deposited InP having exceptional optoelectronic properties (carrier concentration $n_{77K}$ =

$5 \times 10^{13}$ cm$^{-3}$ and mobility $\mu_{77K}$ 3.05 x $10^5$ cm$^2$ Vs$^{-1}$). Interestingly it was found that very pure InMe$_3$ does not undergo the premature reaction with PH$_3$ alluded to earlier. Similar enhancements in purity have been achieved for AlMe$_3$[8] and GaMe$_3$[9] using the adduct process.

The diphos story is a considerable success for collaborative research between university and industrial laboratories in an interdisciplinary programme supported under the Joint Opto Electronic Research Scheme (JOERS; with SERC and DTI funds).

The hazardous nature of PH$_3$ and AsH$_3$ was referred to earlier and alternative precursors are being sought. Thus organophosphines RPH$_2$ or organoarsines RAsH$_2$ might be more amenable as moderately volatile liquids with lower toxicity than the trihydrides. In an attempt to circumvent the problem of the high carbon-phosphorus bond strength, Stringfellow et al.[10] have used mono-tert.-butyl phosphine Bu$^t$PH$_2$ which is believed to decompose by alkene elimination.

$$Bu^tPH_2 \rightarrow PH_3 + C_4H_8 \quad \text{----} \quad (2)$$

The organo arsenic compounds decompose much more readily than organophosphorus compounds and we have shown that phenylarsine PhAsH$_2$ can be used in the decomposition of Ga As.[11] This compound can be synthesized from the readily available phenyl arsonic acid.

$$PhAsO(OH)_2 \xrightarrow{Zn/HCl} PhAsH_2 \quad \text{----} \quad (3)$$

Another approach to these problems is offered by very low pressure MOCVD (sometimes referred to as MOMBE) which is carried out at ca. $10^{-3}$ torr in the absence of a carrier gas. Less volatile precursors can be used under these conditions and we have shown that single source precursors containing the covalently bonded Group III and Group V elements can be used.[12] For example (Me$_2$InPBu$^t_2$)$_2$ will deposit InP at a temperature (<400°C) low enough to avoid the need for excess phosphorus. Similarly Cowley et al.[13] have proposed the use of (Me$_2$GaAsBu$^t_2$)$_2$ as a single source precursor for GaAs deposition and we have deposited AlN, which is an insulator having high thermal conductivity, from $(Me_2AlNPr_2^i)_2$.[14]

Turning now to the II/VI semiconductors it is noteworthy that a range of such materials extending from wide band gaps to narrow band gaps may be deposited by the MOCVD technique.[3] Recent work has been concerned particularly with the development of low temperature processes and the elimination of premature reactions between the Group II metal dialkyl and the Group VI element precursor.[15] One approach has involved the replacement of $H_2S$ and $H_2Se$ by organo-derivatives such as the dialkyls[16] or heterocyclic compounds such as furan $(C_4H_4O)$, $C_4H_4S$ (thiophene) and $C_4H_4Se$ (selenophene)[17], although somewhat higher decomposition temperatures may arise which can cause difficulties[18]. Another successful approach has utilised the fact that dialkyl zinc adducts such as $ZnMe_2(NEt_3)_2$ do not undergo premature reaction with $H_2Se$, and being easier to purify than the dialkyl zinc, produce very high quality ZnSe at relatively low temperatures (200 - 350°C).[19] Similarly it has been demonstrated that adducts of $CdMe_2$ prevent premature reactions with $H_2S$ or $H_2Se$ and lead to high quality epitaxial growth of CdS, CdSe and $CdS_xSe_{(1-x)}$ at 300-500°C.[20] The use of di-isopropyl telluride as the tellurium precursor has led to the deposition of excellent $Cd_xHg_{1-x}Te$ (CMT).[21] Returning to the purification of the dialkyl zinc and dialkyl cadmium precursors Cole-Hamilton et al.[22] have shown that the non-volatile nitrogenous bases such as 3,3'-bipyridyl, p-dimethylamino-benzene and p-dimethylamino pyridine, are suitable for adduct purification of the metal dialkyls analogous to the diphos adduct purification of the Group III metal alkyls.

Another requirement for volatile metallo-organic precursors is the need for suitable dopants for the III-V and II-VI semiconductors. A very low concentration ($10^{15}$-$10^{17} cm^{-3}$) of dopant is normally required but it must be amenable to strict control. For example zinc is used as a p-dopant in GaAs and recently it has been reported[23] that N,N,N',N'-tetramethylethylene diamine $(Me_2NCH_2CH_2NMe_2)$ forms a volatile chelate with dimethyl zinc $ZnMe_2$.TMED which is superior as a dopant to the dialkyl zincs alone.

Thus it is clear that the MOCVD technique for the

epitaxial deposition of semiconductors has generated a demand for a range of high purity fine chemicals and with the need for novel precursors for photo-assisted deposition this demand is bound to increase in the future.

## 3 METAL OXIDES

The metal alkoxides $[M(OR)_x]_n$ (where M is an x-valent metal, R an alkyl group, and n the degree of polymerization) are metallo-organic compounds which are attractive as precursors for the deposition of metal oxides by MOCVD or sol-gel processes.

Although some metal alkoxides had been known for many years, it was not until the 1950's that systematic studies on the relationship between the degree of polymerization (n) and the steric effect of the alkyl groups led to our present understanding of the factors controlling the volatility of the metal alkoxides and culminated in the first monograph on these compounds in 1978.[24]

The synthetic procedures and physico-chemical properties of the metal alkoxides are well documented[24] and will not be dealt with here. Suffice it to say that the use of the bulky tertiary alkoxide groups (e.g. $Bu^tO$) led to the formation of mononuclear volatile compounds of a large number of metals (e.g. Ti, V, Cr, Zr, Nb, Mo, Hf, Ta, W, U). These compounds can be purified by distillation under reduced pressure and being non-corrosive may be kept indefinitely in sealed glass containers without risk of contamination. The main problem in handling metal alkoxides is their ease of hydrolysis even with very low concentrations of water.

$$2M(OR)_x + H_2O \longrightarrow M_2O_x + 2xROH$$

Thus stringent precautions are required to prevent hydrolysis.

However, this property is turned to good account in the use of metal alkoxides to form metal oxides by MOCVD and the Sol-gel process.[25,26] The MOCVD process requires volatile metal alkoxides and is especially suitable for depositing thin films of metal oxides. In the Sol-gel process solubility of the metal alkoxide in a suitable organic solvent is all that is required and this technique is well suited to the bulk preparation of metal oxides as ceramics or glasses.

Some examples of the use of metal oxides are given in Table 2.

## Table 2        Uses of Metal Oxides

| Metal oxide | Uses |
|---|---|
| $Al_2O_3, SiO_2, SnO_2,$ | Catalysts, Anti-reflective coatings, |
| $TiO_2, CrO_2, Fe_2O_3,$ | Anti-static coatings, Insulating films, Transparent conductors, |
| $Y_2O_3, ZrO_2, Nb_2O_5,$ | Ferroelectrics, Dielectrics, Electro- |
| $HfO_2,$ | optic ceramics, Piezoelectric ceramics, |
| $LiNbO_3, LiTaO_3,$ | Photo-refractive materials, Phosphors, |
| $KNbO_3$ | Magnetic storage materials, High |
| $KNb_{1-x}Ta_xO_3, BaTiO_3$ | temperature superconductors. |
| $Ba_2NaNb_5O_{15},$ | |
| $Ba_{1-x}Sr_x\ Na_2O_6,$ | |
| $Bi_{12}SiO_{20}, B_{12}GeO_{20},$ | |
| $Pb_{(1-x)}La_x(Ti_y, Zr_z)O_3,$ | |
| $YBa_2Cu_3O_{7-\delta}.$ | |

Some of the materials listed are heterometal oxides containing two or more different metals and these may pose special problems for the MOCVD technique. Thus, although heterometal alkoxides $M_xM'_y(OR)_z$ are well known[24] they do not necessarily volatilise congruently or have the correct proportions of the different metals. This is not a problem with the sol-gel technique provided that the component metal alkoxides are all soluble in the same solvent and hydrolyse at comparable rates.

As alternatives to metal alkoxides the $\beta$-diketonate derivatives $M(RCOCHCOR')_x$ may be used as volatile precursors for metal oxide deposition and the vapour pressures of a number of metal complexes of the 2,2,6,6-tetramethyl heptane-3,5-dionato ligand (R=R'=Bu$^t$) have been measured.[27] Further enhancement of the volatility of yttrium alkoxides has been achieved by using fluorinated tertiary alkoxide groups such as $(CF_3)_2(CH_3)CO$ taking advantage of the lower intermolecular attractive forces between perfluoroalkyl groups compared to hydrocarbyl groups.[26] A striking example of this effect is shown in the relatively high volatility of the perfluoro-tertiary-butoxides of lithium, sodium and potassium, with the tetrameric potassium derivative $[KOC(CF_3)_3]_4$ subliming at $142°C/0.2$mm Hg.[28] Fluorinated β-diketonates e.g. $CF_3COCHCOCF_3$ also produce enhanced volatility in their metal derivatives but they are liable to produce metal fluorides by thermolysis unless extra oxygen is supplied in the form of water vapour or dioxygen. The barium derivative of the 1,1,1,2,2,3,3-heptafluoro-7,7-dimethyl octane-4,6-dionate ligand ($C_3F_7COCHCOBu^t$; fod) has been used in the MOCVD deposition of $YBa_2Cu_3O_{7-\delta}$[29] and the trifluoroacetates of Y, Ba and Cu were used as precursors in a spray-on process[30].

## 4 METAL FLUORIDES

Nearly 25 years ago Kao and Hockman[31] proposed the use of silicate glass fibres as optical waveguides for telecommunication. Initially the attenuation (>300 dB kM$^{-1}$) due to absorption of light by impurities and Rayleigh scattering restricted transmission to short distances but over the years improvements in purification have reduced the attenuation to values near the limiting range 0.1 - 0.2 dB km$^{-1}$ (at 1.44μm) which is the theoretical value for Rayleigh scattering by silicate glass. This has made long distance optical fibre communication a practical proposition and commercial developments are now well advanced. More recently attention has been directed to the advantages of using fluoride glasses for fibre optics due to the theoretical lower limit of attenuation due to Rayleigh scattering (0.001 dB Km$^{-1}$ at 2.55μm).[32] Multicomponent glasses are being investigated using up to six different metals selected from the following wide range:-
Li, Na, K, Be, Mg, Ca, Sr, Ba, Zn, Cd, Pb, Al, Ga, In, Sc, lanthanides, certain transition metals, Zr, Hf, Th and U. Certain transition metals (Fe, Co, Ni, Cu) and lanthanides (Pr, Nd, Sm, Eu, Tb) are undesirable impurities due to absorption and must be reduced to parts per billion levels to achieve the limiting Rayleigh scattering of the fluoride glass. Also oxygen in the forms of $O^{2-}$ or $OH^-$ is extremely deleterious.

Clearly there are considerable challenges to

inorganic chemists in producing high purity fluoride glasses and chemical vapour purification offers considerable scope[32]. For example oxygen-free $ZrF_4$, which is a major component of most fluoride glasses, can be obtained in a two stage process starting from very pure metal.

$$Zr + 2Cl_2 \rightarrow ZrCl_4$$

$$ZrCl_4 + 4HF \rightarrow ZrF_4 + 4HCl.$$

The use of the volatile fluorinated β-diketonates (e.g. $CF_3COCHCOCF_3$) of the metals has been proposed as precursors for depositing metal fluoride glasses.[33]

Films of metal fluorides are also of interest to the electronics industry and we have shown that the anhydrous calcium derivative $Ca(CF_3COCHCOCF_3)_2$ is a suitable precursor for photo-assisted deposition of $CaF_2$ on GaAs at fairly low temperature (100°C).[34]

It appears therefore that precursors for metal fluorides may constitute a growth area in the fine chemicals area.

REFERENCES

1. H.M. Manasevit, *Appl. Phys. Lett.,* 1968, *12,* 156; H.M. Manasevit and W.I.Simpson, *J. Electrochem. Soc.,* 1969, *116,* 1725.
2. D.C. Bradley, *New Scientist,* 1988, *1608,* 38.
3. A.C. Jones, *Chemtronics,* 1989, *4,* 15; and references therein.
4. R.H. Moss and J.S. Evans, *J. Cryst. Growth,* 1981, *55,* 129; M.M. Faktor, A.K. Chatterjee, R.H. Moss and E.A.D. White, *J. Physique,* 1982, *43,* C5; S.J. Bass, M.S. Skolnick, H. Chudzynska and L.M. Smith, *J. Cryst. Growth,* 1986, *75,* 221.
5. D.C.Bradley, H. Chudzynska, M.M. Faktor, D.M. Frigo, M.B. Hursthouse, B. Hussain and L.M. Smith, *Polyhedron,* 1988, *7,* 1289.
6. D.C. Bradley, H. Chudzynska and M.M. Faktor, PCT WO85/04405, 26th March, 1985.
7. E.J. Thrush, C.G. Cureton, and A.T.R. Briggs, *J. Cryst. Growth,* 1988, *93,* 870.

8. A.C. Jones, P.R. Jacobs, R. Cafferty, M.D. Scott, A.H. Moore, and P.J. Wright, J. Cryst. Growth, 1986, 77, 47.
9. A.C. Jones, G. Wales, P.J. Wright and P.E. Oliver, Chemtronics, 1987 2, 83.
10. C.H. Chen, C.A. Larsen, G.B. Stringfellow, D.W. Brown, and A.J. Robertson, J. Cryst. Growth, 1986, 77, 11.
11. R.D. Hoare, O.F.Z. Khan, J.O. Williams, D.C. Bradley, D.M. Frigo, H. Chudzynska, P. Jacobs, A.C. Jones and S.A. Rushworth, Chemtronics, 1989, 4, 78.
12. D.A. Andrews, D.C. Bradley, G.J. Davies, M.M. Faktor, D.M. Frigo and E.A.D. White, Semicond. Sci. Technol., 1988, 3, 1053; G.J. Davies and D.A. Andrews, Chemtronics, 1988, 3, 3.
13. A.H. Cowley, B.L. Benac, J.G. Ekerdt, R.A. Jones, K.B. Kidd, J.Y. Lee and J.E. Miller, J. Am. Chem. Soc., 1988, 110, 6248.
14. D.C. Bradley, D.M. Frigo and E.A.D. White, UK Patent Application Nos. 8804707 and 8805524.
15. L.M. Smith and J. Thompson, Chemtronics, 1989, 4, 60.
16. H. Mitsuhashi, H. Mitsuishi and H. Kukimoto, J. Appl. Phys., 1985, 24, 1864.
17. P.J. Wright, R.J.M. Griffiths, B. Cockayne, J. Cryst. Growth, 1984, 66, 26.
18. B. Cockayne, P.J. Wright, M.S. Skolnick, A.D. Pitt, J.O. Williams and T.L. Ng, J. Cryst. Growth, 1985, 72, 17.
19. B. Cockayne, P.J. Wright, A.J. Armstrong, A.C. Jones and E.D. Orrell, J. Cryst. Growth, 1988, 91, 57; P.J. Wright, P.J. Parbrook, B. Cockayne, A.C. Jones, E.D. Orrell, K.P. O'Donnell, and B. Henderson, J. Cryst. Growth, 1989, 94, 441.
20. P.J. Wright, B. Cockayne, A.C. Jones, E.D. Orrell, P. O'Brien and O.F.Z. Khan, J. Cryst. Growth, 1989, 94, 97.
21. J. Thompson, P. Mackett, L.M. Smith, D.J. Cole-Hamilton, and D.V. Shenai-Khatkhate, J. Cryst. Growth, 1988, 86, 233.
22. D.V. Shenai-Khatkhate, E.D. Orrell, J.B. Mullin, D.C. Cupestino and D.J. Cole-Hamilton, J. Cryst. Growth, 1986, 77, 27.
23. P.J. Wright, B. Cockayne, A.C. Jones and E.D. Orrell, J. Cryst. Growth, 1988, 91, 63.
24. D.C. Bradley, R.C. Mehrotra, and D.P. Gaur, Metal Alkoxides, Academic Press, London (1978).
25. L.G. Hubert-Pfalzgraf, Nouv. J. Chem. 1987, 11, 663.
26. D.C. Bradley, Chem. Rev., 1989, 89, 1317; Phil. Trans. Roy. Soc. Lond. A, 1990, 330, 167.
27. H.R. Brunner and B.J. Curtis, J. Therm. Anal., 1973,

**5**, 111.
28. R.E.A. Dear, F.W. Fox, R.J. Fredericks, G.E. Gilbert, and D.K. Huggins, Inorg. Chem., 1970, **9**, 2590.
29. Jing Zhao, Klaus-Hermann Dahmen, H.O. Maraj, L.M. Tonge, T.J. Marks, B.W. Wessels and C.R. Kannewurf, Appl. Phys. Lett., 1988, **53**, 1750.
30. A. Gupta, R. Jagannathan, E.I. Cooper, E.A. Giess, J.I. Landman, and B.W. Hussey, Appl. Phys. Lett., 1988, **52**, 2077.
31. K.C. Kao and G.A. Hockman, Proc. Inst. Electr. Eng. 1966, **113**, 1151.
32. A.E. Comyns, Chem. in Britain, 1986, 47; P.W. France, S.F. Carter, M.W. Moore and C.R. Day, Br. Telecom. Technol. J., 1987, **5**, 28.
33. J.M. Power and A. Sarhangi, U.S. Patent 4, 718, 929, Jan. 12th, 1988.
34. A.W. Vere, K.J. Mackey, D.C. Rodway, P.C. Smith, D.C. Bradley and D.M. Frigo, Advanced Mat., 1989, **11**, 399.

**Rare Earth Materials**

Rare Earths in the Electronics Industry
K. A. Gschneidner Jr (Ames Laboratory, University of Iowa, USA)

# Rare Earths in the Electronics Industry

K. A. Gschneidner, Jr.

AMES LABORATORY* AND DEPARTMENT OF MATERIALS SCIENCE AND
ENGINEERING, IOWA STATE UNIVERSITY, AMES, IOWA 50011, USA

## 1 INTRODUCTION

The rare earth elements include the Group IIIA elements Sc, Y and the lanthanide elements La, Ce, Pr, Nd, Pm, Sm, Eu, Gd, Tb, Dy, Ho, Er, Tm, Yb and Lu. This definition, which is consistent with that of the International Union of Pure and Applied Chemistry, will be used throughout this paper. It is noted that many scientists and engineers use the term "rare earths" to mean the 15 lanthanide elements and they do not consider Sc and Y to be "rare earths".

The lanthanide group is subdivided into the "light lanthanides", which refers to the first half of the group (La through Eu), and the "heavy lanthanides", which are the last members of the group (Gd through Lu), where the terms "light" and "heavy" refer to the element's atomic weight. These terms come from the early history of the rare earths, where it was noted that many minerals have an abundance of the light elements, and only a few that have abundances which are high in the heavy lanthanides. Y is always found, usu-

---

*Operated for the U. S. Department of Energy by Iowa State University under contract no. W-7405-ENG-82. This work was supported by the Office of Basic Energy Sciences.

ally in large proportions, in the "heavy" group, while Sc does not normally occur in the rare earth minerals or ores.

## 2 ELECTRONIC CONFIGURATIONS

The electronic configurations of the rare earths are important and govern their chemistry, metallurgy and physics, and account for their interesting optical and magnetic properties and applications. The neutral (gaseous) atom configurations are shown on the left side of Figure 1. These electron configurations are not too important because we are primarily concerned with the solids; however, they have an impact on the vapor pressures and boiling points of the rare earth metals. The more important configurations are those shown in the middle and on the right side of Figure 1. The most important oxidation states (the middle set of columns) are the trivalent series ($M^{3+}$), where the number of f electrons in the lanthanide series varies from 0 for La to 14 for Lu. There are several elements

### Electronic Configurations of the Rare Earths

| Element | Neutral Atom Configuration | | 4f Configuration of Known Oxidation States | | | Metallic State No. of Electrons | |
|---|---|---|---|---|---|---|---|
| | | | $M^{2+}$ | $M^{3+}$ | $M^{4+}$ | Valence | 4f |
| Sc | | $3d4s^2$ | - | 0 | - | 3 | 0 |
| Y  | | $4d5s^2$ | - | 0 | - | 3 | 0 |
| La | | $5d6s^2$ | - | 0 | - | 3 | 0 |
| Ce | $4f$ | $5d6s^2$ | - | 1 | 0 | 3 | 1 |
| Pr | $4f^3$ | $6s^2$ | - | 2 | 1 | 3 | 2 |
| Nd | $4f^4$ | $6s^2$ | - | 3 | - | 3 | 3 |
| Pm | $4f^5$ | $6s^2$ | - | 4 | - | 3 | 4 |
| Sm | $4f^6$ | $6s^2$ | 6 | 5 | - | 3 | 5 |
| Eu | $4f^7$ | $6s^2$ | 7 | 6 | - | 2 | 7 |
| Gd | $4f^7$ | $5d6s^2$ | - | 7 | - | 3 | 7 |
| Tb | $4f^9$ | $6s^2$ | - | 8 | 7 | 3 | 8 |
| Dy | $4f^{10}$ | $6s^2$ | - | 9 | - | 3 | 9 |
| Ho | $4f^{11}$ | $6s^2$ | - | 10 | - | 3 | 10 |
| Er | $4f^{12}$ | $6s^2$ | - | 11 | - | 3 | 11 |
| Tm | $4f^{13}$ | $6s^2$ | - | 12 | - | 3 | 12 |
| Yb | $4f^{14}$ | $6s^2$ | 14 | 13 | - | 2 | 14 |
| Lu | $4f^{14}$ | $5d6s^2$ | - | 14 | - | 3 | 14 |

**Figure 1**  Electronic configurations of the rare earth elements

which change oxidation states. One of those is Ce, which can go from the trivalent state with one f electron to a tetravalent state with no f electrons. Similarly trivalent Pr which has two f electrons gives up one of them when it becomes tetravalent. Tb, which normally has eight f electrons in its trivalent state, has seven (a half-filled shell) when it becomes tetravalent. There are also three elements which have a tendency to be divalent in addition to their normal trivalent state. These are Sm, Eu, and Yb each of which gains a 4f electron in the divalent state to have six, seven and fourteen f electrons, respectively. In the case of Sm and Eu there is a tendency toward a half-filled shell (seven 4f electrons), and for Yb the tendency toward a completely filled 4f shell. These valence changes occur because of the extra stability that an element gains when it has a completely empty 4f level (i.e. tetravalent Ce), a half-filled level (i.e. divalent Eu and tetravalent Tb), or completely filled 4f level (i.e. divalent Yb).

These tendencies also show up in the metallic state, see the two columns on the right side of Figure 1, the electronic configurations of the rare earth metals. It is seen that Eu and Yb are divalent with a half-filled shell and completely filled shell, respectively, while the remaining rare earth metals are trivalent.

The influence of these anomalous valence states shows up in the chemical, metallurgical and physical properties, and one must be aware of these anomalies in working with and using these elements.

## 3 ABUNDANCES

Of the 83 naturally occurring elements the 16 naturally occurring rare earth elements fall into the 50th percentile of the elemental abundances. Cerium, which is the most abundant, ranks 28th and Tm, the least abundant, ranks 63rd. Collectively, if the rare earths were considered as one element, they would rank as the

22nd most abundant (at the 75th percentile). As a whole, the heavy lanthanides, Gd to Lu, are less abundant than the light lanthanides, La to Eu, and thus are usually more expensive. Furthermore, the even atomic number metals (Ce, Nd, Sm, Gd, Dy, Er and Yb) are more abundant than their neighboring odd atomic number elements (La, Pr, Pm [which is radioactive], Eu, Tb, Ho, Tm and Lu). These variations are evident in the prices of the individual elements and their compounds.

Many of the common elements (such as Hg, Cd, I and Se) are less abundant than the rare earth elements, and several more (such as W, Mo, Sn, Ge and Pb) are equally abundant. Thus we note that the term "rare" is a misnomer when referring to these elements, but common usage prevails.

The rare earth reserves for individual countries are given in Table 1. Countries for which the reserves are less than 100,000 metric tons REO are not listed in

Table 1    Rare earth reserves (metric tons of rare earth oxide)

| Country | Reserves REO |
|---|---|
| Australia | 755,000 |
| Brazil | 20,000 |
| Canada | 182,000 |
| China | 36,000,000 |
| Egypt | 100,000 |
| India | 2,200,000 |
| Malawi | 297,000 |
| Malaysia | 30,000 |
| Republic of South Africa | 357,000 |
| United States | 5,500,000 |
| U.S.S.R. | 450,000 |
| Others[a] | 172,000 |
| Total | 45,964,000 |

[a]Europe 50,000; Africa - 63,000; Asia - 59,000

Table 1 unless they make substantial contributions to the current world production. This is the situation for Brazil and Malaysia, which account for 2 and 7%, respectively, of the 1989 world's production. Countries which have large reserves (>100,000 tons REO) but do not contribute to the world's mining production (< 1%) are: Canada, Egypt, Malawi and the Republic of South Africa.

## 4 MINERALS AND ORES

Of the ~160 minerals known to contain rare earths, the three most important commercial ones are bastnasite, monazite and xenotime. The first two are sources of the light lanthanides and account for about 96% of the rare earths being utilized, while the last is a source of the heavy lanthanides and Y. Other minerals which have been mined for their rare earth content are apatite, euxenite, and gadolinite. Allanite, fluorite, perovskite, sphene and zircon are potential future sources. Of the eight minerals, most are processed for other constituents and the rare earths may be extracted as a by-product but, because of economics, the rare earths generally are not removed when these ores are mined and processed. The major exception is found in the USSR where the rare earths are extracted from apatite. The idealized chemical compositions of these eleven minerals are given in Table 2.

<u>Table 2</u>  Composition of some of the more important rare earth minerals (after Gschneidner[1])

| Name | Idealized composition | Primary rare earth content |
|---|---|---|
| allanite | $(Ca,Fe^{2+})_2(R,Al,Re^{3+})_3Si_3O_{13}H$ | R = lights |
| apatite | $Ca_5(PO_4)_3F$ | R = lights |
| bastnasite | $RCO_3F$ | R = lights (60-70%) |
| euxenite | $R(Nb,Ta)TiO_6 \cdot xH_2O$ | R = heavies (15-43%) |
| fluorite | $CaF_2$ | R = heavies |
| gadolinite | $R_2(Fe^{2+},Be)_3Si_2O_{10}$ | R - heavies (34-65%) |
| monazite | $RPO_4$ | R = lights (50-78%) |
| perovskite | $CaTiO_4$ | R = lights |
| sphene | $CaTiSiO_4X_2 (X = \frac{1}{2}O^{2-}, OH^- \text{ or } F^-)$ | R = lights |
| xenotime | $RPO_4$ | R = heavies (54-65%) |
| zircon | $ZrSiO_2$ | R = either |

In addition to the three major minerals, there are two other important sources of the heavy lanthanides and Y. One of these is the U tailings in the Elliot Lake region of Ontario, Canada. This new source has boosted the free-world's Y production by about 25%. The second source is the Longnan clays in the southern end of Jiangxi province in southeast China. This ionic absorption clay is unique and accounts for ~2% of China's rare earth production.

The rare earth distribution in the three commercial minerals from various countries and the two unique sources are given in Table 3. The materials presented in this table account for more than 90% of the current rare earth market.

## Bastnasite

Bastnasite, which is a fluorocarbonate, accounts for 54% of the rare earths utilized today. About 40% comes from the Molycorp's open pit mine at Mountain Pass, California about 75 miles from Las Vegas. The remaining 14% comes from the Baiyunebo iron ore mine, about 85 miles from Baotou, Inner Mongolia. The Baiyunebo mine is the largest rare earth deposit in the world and alone accounts for 70% of the known world's reserves. However, the rare earths are not present entirely as bastnasite - the mineral product is mainly an iron ore, containing bastnasite and monazite (in a 3:1 to 2:1 ratio of bastnasite:monazite). In addition, there is a rock overburden, consisting primarily of CaO and MgO, plus Nb and rare earths, which is just set aside and stored at the mine site.

Examination of Table 3 shows that the Chinese deposit, relative to the Mountain Pass deposit, is slightly richer in the heavy lanthanides and Y, primarily at the expense of the La content, which is about 10% greater in the California bastnasite. Other significant bastnasite deposits are found in Burundi, Sweden and New Mexico in the U.S.A., but none are currently being mined.

Table 3  Rare earth content of several primary source minerals (after Gschneidner[1], Zhang et al.[2], Li[3] and Anstett[4])

| Rare earth element | Bastnasite Calif. (%) | Bastnasite China[a] (%) | Monazite China[b] (%) | Monazite Australia[c] (%) | Monazite Brazil and India (%) | Xenotime Malaysia (%) | Uranium residues Ontario, Canada (%) | Longnan clay China (%) |
|---|---|---|---|---|---|---|---|---|
| La | 32.0 | 22.8 | 23.4 | 23.9 | 22.8 | 0.5 | 0.8 | 2.2 |
| Ce | 49.0 | 49.8 | 45.7 | 46.0 | 45.7 | 5.0 | 3.7 | 1.1 |
| Pr | 4.4 | 6.2 | 4.2 | 5.0 | 5.0 | 0.7 | 1.0 | 1.1 |
| Nd | 13.5 | 18.5 | 15.7 | 17.4 | 18.9 | 2.2 | 4.1 | 3.5 |
| Sm | 0.5 | 1.0 | 3.0 | 2.5 | 3.0 | 1.9 | 4.5 | 2.3 |
| Eu | 0.1 | 0.2 | 0.1 | 0.05 | 0.1 | 0.2 | 0.2 | 0.1 |
| Gd | 0.3 | 0.7 | 2.0 | 1.5 | 1.7 | 4.0 | 8.5 | 5.7 |
| Tb | | 0.1 | 0.1 | 0.04 | 0.2 | 1.0 | 1.2 | 1.1 |
| Dy | | 0.1 | 1.0 | 1.2 | 0.5 | 8.7 | 11.2 | 7.5 |
| Ho | 0.1 | | 0.1 | 0.05 | 0.1 | 2.1 | 2.6 | 1.6 |
| Er | | | 0.5 | 0.2 | 0.1 | 5.4 | 5.5 | 4.3 |
| Tm | | 0.1 | 0.5 | 0.01 | – | 0.9 | 0.9 | 0.6 |
| Yb | | | 0.5 | 0.1 | 0.1 | 6.2 | 4.0 | 3.3 |
| Lu | | | 0.1 | 0.04 | – | 0.4 | 0.4 | 0.5 |
| Y | 0.1 | 0.5 | 3.0 | 2.4 | 2.0 | 60.8 | 51.4 | 64.1 |

[a] Baiyunebo iron ore mine, Inner Mongolia

[b] Guangdong/Guangxi Provinces

[c] Western Australia

One of the major advantages of bastnasite is that it contains extremely small amounts of Th (<0.02%) and does not have the radioactive environmental problems associated with it that monazite does (see below).

**Monazite**

Monazite is the most important ore source for the rare earths and accounts for about 42% of the free-world utilization. Monazite is a rare earth phosphate containing 5 to 10% $ThO_2$ which presents some interesting problems associated with its processing and utilization. Because Th is a fissionable material, several countries, notably Brazil and India, remove and stockpile the Th as a strategic item, and the Th-free monazite is exported. Australia and Malaysia have no such export restrictions. Because of Th's natural radioactivity, most countries, notably Australia and the U.S.A., have stringent environmental requirements on the Th content of the monazite processed and utilized therein. The five major countries producing monazite are (the number in parentheses after each country is their percentage of the total world monazite production in 1989): Australia (34%), Brazil (5%), China (31%), India (11%) and Malaysia (14%). Other countries mining monazite are the Republic of Korea, Sri Lanka, Thailand, U.S.A., U.S.S.R. and Zaire but the combined total was about 1000 metric tons in 1989.

The rare earth distribution in three different monazites is given in Table 3, where it is noted that the monazites have a significantly larger concentration of heavy lanthanides plus Y than the bastnasite. The Chinese and Australian monazites have nearly the same rare earth distribution, the Chinese being slightly higher in the heavy lanthanides and Y. Both of these ore sources are richer in the heavy lanthanides and Y than the Brazilian and Indian monazites which have almost identical rare earth distributions.

Most commercial monazite sources are alluvial and beach sand deposits. Monazite is a minor component (1 to 20%) of the heavy mineral sands whose primary constituents are rutile, ilmenite, cassiterite and ziron. In the past, these sands were primarily processed for their Ti, Zr or Sn content, and thus the monazite production varied as the markets for Ti, Zr and Sn have gone up or down. However, the demand for rare earths has grown sufficiently in the last ten years that some of the heavy mineral sands can be processed economically for their monazite and the excess Ti or Zr or Sn is stored when these markets are down.

## Xenotime

Xenotime is the major ore source for the heavy lanthanides and Y. It is a phosphate containing up to ~3% each of $U_3O_8$ and $ThO_2$. The rare earth distribution is shown in Table 3, and the contrasts between xenotime and the bastnasites and monazites are quite evident - the light lanthanide concentrations are smaller by a factor of about ten, while the heavy lanthanide and Y concentrations are up by a factor of ten to several hundred. Because of high Y and heavy lanthanide concentrations, most of the xenotime is used as a source material for the individual rare earth elements rather than being used as a mixture of heavy rare earths. The major producers of xenotime are Australia, China and Malaysia, while deposits are reported to exist in the United States, Norway and Brazil.

## Other Important Minerals

The Longnan clay and the U residues are also important sources of the heavy lanthanides and Y, the former having higher concentrations of these elements than xenotime, while the latter having somewhat lower concentrations than xenotime, see Table 3. Although these two sources account for only a small fraction of the total consumption of rare earths today, they will play an increasingly important role in the future.

Many of the rare earth deposits in Jiangxi province are of the so-called "ion adsorption" type. These are weathered granites, in which the rare earth ions are adsorbed on the surface of Al silicates, such as kaolin. There are two basic types of minerals found in this province: one rich in the light lanthanides (~75%) known as the Xunwu mineral, and the other rich in the heavy lanthanides and Y known as the Longnan mineral (see Table 3). These ore bodies are easily mined because these are in the form of sand and powder. In addition to this clay large monazite, xenotime and gadolinite deposits are also found in Jiangxi province.

## 5 BENEFICIATION OF ORES

The processes described below are those utilized by the largest rare earth producers, namely the Molycorp process for bastnasite at Mountain Pass[5]; the Rhône-Poulenc process for monazite obtained from a variety of sources throughout the world[6]; the Yue Long (Shanghai) process for monazite[2] and the Baotou process for the combined bastnasite/monazite ore. Other producers may use, for upgrading their ores, processes that are similar to those noted below, or they can use alternate methods similar to those described in the monograph by Callow[7] on rare earth industrial processes.

### Bastnasite

The California bastnasite is mined by open pit methods. The ore which contains ~8% rare earth oxide (REO), is ground to a powder and is upgraded to a product containing 60% REO by a hot froth flotation technique. The heavier barite ($BaSO_4$) and celestite ($SrSO_4$) settle out while the bastnasite and other light minerals are floated-off. The 60% concentrate is treated with 10% HCl to dissolve the calcite ($CaCO_3$), up-grading the REO content to 70%.

The 70% REO concentrate is roasted to convert $Ce^{3+}$ to $Ce^{4+}$. After cooling, the material is leached with HCl dissolving the trivalent lanthanides (La, Pr, Nd, Sm, Eu and Gd) leaving behind the Ce concentrate which is refined to various grades and marketed. The dissolved lanthanides are separated into two groups, La-Pr-Nd and Sm-Eu-Gd, by solvent extraction. The La, Pr, Nd concentrate may be marketed as a Ce-free rare earth product or further processed for the individual elements by continued solvent extraction processes, as shown in Figure 2.

The Sm-Eu-Gd fraction is treated with a reducing agent to reduce Eu to the divalent state while the $Sm^{3+}$ and the $Gd^{3+}$ remain in the higher oxidation state. Divalent Eu is precipitated as the insoluble $EuSO_4$, oxidized to $Eu^{3+}$, recovered and sold as $Eu_2O_3$. The Sm and Gd are separated by a solvent extraction step.

Little or no details are available concerning the processing of the Baiyunebo ore, and the information given below has been gleaned from little bits and pieces from a variety of sources. As noted above, the Baiyunebo ore is primarily mined for its Fe content and shipped ~180 km to Baotou by rail for processing. Little or no processing is done at the mine because of the lack of water; Baotou has plentiful water since it is located on the Yellow River.

A schematic outline of the ore dressing and bastnasite processing of this complex ore is shown in Figure 3. The initial steps of grinding and magnetic separation removes the rare earth oxide by-product from the iron concentrate which is processed to various iron and steel products. The rare earth residue is upgraded initially from ~10% to ~30% by gravity concentration and then up to 68% by multi-stage flotation process. At this stage, the monazite is separated by another flotation method using a specially prepared collector to separate the monazite from the bastnasite. The bastnasite is cracked by heating in sulfuric acid at 400-500°C, and then following a number of physical and chemical steps (Figure 3), is converted to a chloride solution for further processing. The chlorides can be used for making mischmetal, catalysts, or as the feedstock for separation processes to give the individual rare earths or concentrates of them in steps similar to

74  Fine Chemicals for the Electronics Industry II

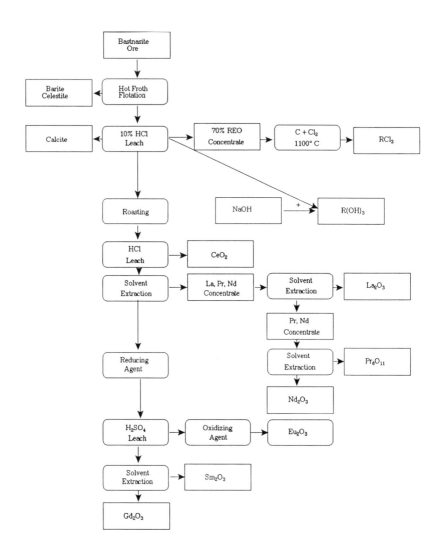

Figure 2  Flow chart of the Molycorp process for the beneficiation of bastnasite and the separation and purification of the rare earth elements

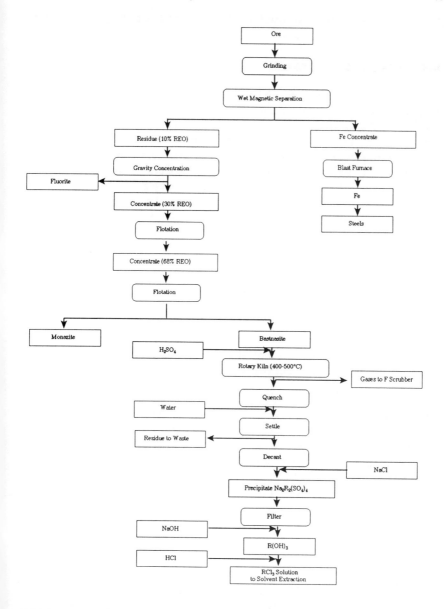

**Figure 3** Flow chart of the process for the beneficiation of the bastnasite/monazite Baiyunebo ore

that outline in the bottom two-thirds of Figure 2, or the lower half of Figure 4.

## Monazite (and Xenotime)

The beach sands which contain monazite and other heavy minerals are recovered by using various techniques, including dredges, and spirals. The sands mainly consist of Ti and Zr minerals and from 1 to 20% monazite (and/or xenotime). This is true of the ores from Australia, Brazil, China and India. The Malaysian monazite/xenotime, however, is a by-product of alluvial tin mining. The monazite (xenotime) is separated from the other minerals by a combination of gravity, electromagnetic and electrostatic techniques, and then is cracked by using either the acid process or the basic process. The basic process used by Rhône-Poulenc for cracking monazite is described below. The Chinese process used at the Yue Long Chemical Plant in Shanghai[2] is quite similar and so it will not be discussed in any detail. For more information on the acid process see the treatise by Callow.[7] Xenotime, which is also rare earth phosphate is usually treated in the same manner as the monazite.

The monazite is initially treated with a 70% NaOH solution and heated in an autoclave at ~150°C for several hours. After cooling the $Na_3PO_4$ is dissolved out by adding water and is sold as a by-product. The rare earths remain behind as insoluble $R(OH)_3$ which still contains the 5 - 10% Th. Two different processes are used by Rhône-Poulenc to remove the Th. One process is to dissolve $R(OH)_3$ in HCl or $HNO_3$ and then selectively precipitate $Th(OH)_4$ by addition of NaOH and/or $NH_4OH$. The second process is to carefully add HCl to $R(OH)_3$ to lower the pH to 3 - 4, and to form the soluble $RCl_3$ and then filter-off the insoluble $Th(OH)_4$. Both processes are shown in the flow sheet, Figure 4.

The Th-free rare earth solution may be converted to a usable form of the mixed elements (e.g. the hydrated chloride, carbonate, hydroxide or oxide), or it

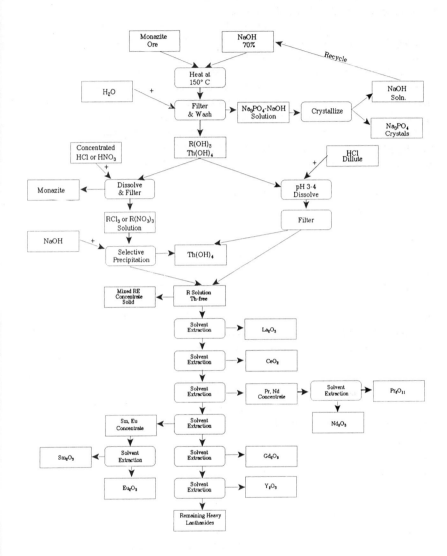

Figure 4  Flow chart of the Rhône-Poulenc basic process for the beneficiation of monazite and the separation and purification of the rare earth elements

may be further processed to obtain other mixed concentrates or the individual elements (Figure 4). Rhône-Poulenc uses only the liquid-liquid solvent extraction technique to separate the rare earth elements, rather than a combined chemical and solvent extraction process as employed by Molycorp for bastnasite (see Figure 2). In the Rhône-Poulenc process, the La is extracted first, then Ce, and successively the other lanthanides, according to their increasing atomic numbers. Y is removed within the heavy lanthanide group and its location with respect to these elements can vary according to the choice of organic extractants.

## 6 SEPARATION CHEMISTRY

Two of the most important spin-offs which resulted from research on the atomic bomb in World War II and the years immediately following were the development of the ion exchange and the liquid-liquid solvent extraction techniques for separating the rare earth elements. Today, the liquid-liquid solvent extraction (more commonly called solvent extraction [SX]) method is used by the major rare earth producers to separate the natural mixtures into the individual elements. This technique is capable of preparing 99.9% pure elements for all of the rare earths and, in the case of Y, a 5 nines or better pure material. On the other hand, the ion-exchange technique has been used to prepare *all* of the rare earth elements at the 5 nines or better purity level.

### Solvent Extraction

In this process, two immiscible or partially immiscible solvents containing dissolved rare earths are mixed, the solutes are allowed to distribute between the two phases until equilibrium is established, and then the two liquids are separated. The concentrations of the solutes in the two phases depend upon the relative affinities for the two solvents, one of which is an aqueous phase (generally a strongly acid solution containing $HCl$, $HClO_4$ or $HNO_3$) and the other an organic phase (such as tributylphosphate). More details can be found in the reviews by Gschneidner[1], Kaczmarek[6] and Powell[8].

## Ion Exchange

In the ion exchange process, a metal ion, $R^{3+}$ in solution, exchanges with protons on a solid ion exchanger - a natural zeolite or a synthetic resin - which is normally called the resin. The tenacity with which the cation is held by the resin depends upon the size of the ion and its charge. However, no separation of the rare earth is possible because the resin is not selective enough. By introducing a complexing agent, $A^*$, separation is possible, if the equilibrium constant, K (also called the stability constant), for the reaction

$$R + 3A^* \rightleftarrows RA_3^*$$

varies sufficiently from one rare earth to another for the separation to take place. The two most important complexing agents are EDTA (ethylene diamine tetra-acetate) and HEDTA (hydroxyethylethylene diamine tri-acetate). More details are given in the reviews by Gschneidner[1] and Powell[8].

To date, the ion exchange process is the best method for producing the highest purity rare earth element. Only for Y is the liquid-liquid extraction process capable of reaching the same purity level obtainable with ion exchange. The major disadvantage of ion exchange is the length of time it takes to purify a given amount. It can take months before the last remaining lanthanide, from a mixture which has been loaded on the columns, is finally eluted from the resin bed. A number of companies throughout the world still use the ion exchange method for separating and purifying the rare earths. A few use it exclusively and others use it in combination with liquid-liquid solvent extraction, whereby the initial separations are made by using the latter technique and if needed, the final purification by using ion exchange.

### New Routes for Separating and Refining

A few years ago, several companies and/or individuals made pronouncements and claims of new methods or big advancements in current technology for separating the rare earth elements. However, in the intervening years, nothing has developed out of all these proclamations, and it is extremely doubtful that any major advance will occur in separation technology of the rare earth elements in the next ten years. The problem still remains the same, except for the dual valence elements (Ce, Sm, Eu and Yb), the chemical properties of the lanthanides are quite similar and change slowly in a regular and systematic manner across the series, and this makes their separation a difficult and tedious process. This is not to say that there will be no improvements in the future. To the contrary, basic and applied research, and new and improved technologies, driven by international competition will lead to more efficient separation and refining processes but by slow and steady steps. A major breakthrough, which would result in a cheap, efficient, large-scale rare earth separation method, is extremely unlikely in the foreseeable future.

### 7 CHEMICAL ANALYSIS AND PURITY OF MATERIALS

The purity required of the rare earth material greatly depends upon the application it is intended for. If one is making phosphors for color TV - the red color from your TV set is due to an Eu activator in an Y-based host - 99.999% pure materials are needed. This is a big and important rare earth market. For lasers (primarily Nd) and other optical uses the rare earths also need to be this pure because if other rare earths are present, they will destroy the spectral purity needed. For lenses and garnets 99.9% pure rare earths are generally pure enough for these markets.

For glass coloring, pigments, other ceramics, decolorizing, condensers and nuclear applications ~99% pure materials will be satisfactory. Permanent magnets, magnetostrictive materials, heat resistance alloys, H-storage and glass polishing materials, and emission catalysts generally do not require such high purities (95 to 99% pure is sufficient), especially with respect to other rare earths.

There are also mixed rare earth products. Generally they contain 95% total rare earths and 5% other non-rare earth materials. These mixed rare earth products are used in steel, ductile iron, magnesium and other metal alloys, cracking catalysts, lighter flints, and carbon arcs.

As noted above the purity requirements for the rare earth materials vary considerably depending on their end uses. But there is a major problem when one discusses chemical analyses - what do the numbers mean? Sale persons generally quote values in weight percent or weight ppm (normally they do not use atomic percent). Furthermore, there is a second problem: when someone claims that their material is 99.999% pure, is it with respect to the rare earth elements only? Or is it with respect to just the metallic elements (including the rare earths)? Or is it with respect to all of the elements in the periodic table? It makes a big difference as to how pure the material really is.

A variety of chemical analysis techniques are used to determine the purity of the rare earth products supplied by the producer. These will vary depending upon nature of the product - a mixture of rare earths, a concentrate of a given element (90 - 99% pure) or a phosphor grade material (99.999% pure). The techniques used are atomic absorption[9], spectrophotometric[10], spark source mass spectrometry[11], x-ray excited optical luminescence[12], neutron activation[13], inductively coupled plasma atomic emission[14], inductively coupled plasma mass spectrometry[14], and laser source mass spectrometry[15]. For analyzing metallic samples vacuum fusion is used to determine H, N and O, and a combusion-chromatographic method to determine the C content.

## 8 APPLICATIONS

An overview of the more important uses of the rare earth elements is summarized in Figure 5. The applications utilizing the mixed rare earths generally involve five of the first six lanthanides. These uses are shown above the horizontal row of lanthanide elements. The "X" over Pm means that this element is not involved in any of the rare earth applications, either in mixtures or individually, because Pm is a radioactive element which does not occur naturally. The use of rare earths as mixtures accounts for 75 to 90% of the total volume of rare earths consumed today and the growth rate for mixtures is about 2% per annum.

The important uses of separated, individual rare earths are listed below the horizontal row of lanthanide elements and to the right of Sc and Y. As seen,

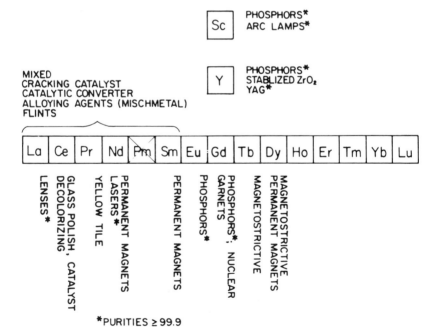

Figure 5   The major uses of the individual and mixed rare earths

there are no major uses involving the last five lanthanides. In addition to utilization in basic and applied research, there are some minor uses for these elements. These include Er as a laser material and Yb metal as a high pressure indicator due to its large resistivity drop (about a factor of ten) at ~4 GPa (~40 kbar).

The use of individual rare earths accounts for ~75% of the monetary value of the world-wide rare earth market (~$350 million), but only 10 to 25% of the tonnage. Furthermore, the growth rate of this segment of the rare earth market is ~15% per year.

Many of the uses shown in Figure 5 are outside of the scope of this Conference and nothing more will be said about them. The remaining applications will be divided up into four groups: optical, magnetic, magneto-optic, and electrical and electronic. A final subsection will discuss possible future applications.

## Optical Applications

Most optical applications require extremely high purity rare earth materials (of the order of 99.999% pure); in some cases even 1 ppm atomic of a certain impurity can have a pronounced affect on the color purity required in the particular application.

The most important use in this category is the use of rare earths as activators in phosphors (in particular Eu, but also Ce, Sm, Tb, Er, Tm) and as the host material (primarily Y compounds, but also Gd). The phosphors are used in CRT's (cathode ray tubes)[16], trichromatic fluorescent tubes[17] and high pressure mercury vapor lamps. The CRT's involve color TV's, radar screens, and avionic and data displays. The red color in TV screens is due to Eu in a Y host material such as $Y_2O_2S$ or $YVO_4$.[16] This is the biggest market in the optical area. The use of rare earth phosphors in trichromatic fluorescent lighting is also important[17] and has shown considerable growth over the last few years. This growth is expected to continue and even increase especially as energy prices continue to rise ~7% per year through the remainder of this century. The trichromatic fluorescent lights are more expensive, but since they are much more efficient (two tubes are as bright as three conventional fluorescent tubes) the operating and initial capital costs are significantly lower.

It should be noted that trivalent Eu emits a red color, but in the divalent state it emits a blue-violet radiation, and some Eu is consumed for its use as a blue phosphor.

Another important use of the rare earths is as x-ray phosphors.[18] Compounds with the lanthanide elements, which have no $4\underline{f}$ electrons (La), or a half filled $4\underline{f}$ level (Gd), are doped with a lanthanide activator (Tb or Tm) and have replaced $CaWO_4$ as x-ray intensifying screens because of their much higher lumin output, thus reducing x-ray exposures by as much as 10 times. Y compounds are not used because of its lower atomic number and thus lower x-ray stopping power as compared with La or Gd. The most commonly used compounds are LaOBr:Tb, LaOBr:Tm and $Gd_2O_2S$:Tb. In addition to lowering patient dosage, there are fewer problems with patient movement thus yielding sharper images, and by using smaller grain film along with the sharper images, more details can be seen in the x-ray films.

The use of rare earths in lasers also requires 5 to 6 nines purities of the rare earth lasing materials. The most important application is the use of Nd and Y in Nd-YAG (yttrium aluminum garnet) lasers, one of the three most popular lasers in use today.[19] Other rare earths that have been used as a lasing ion are Ho, Er and Tm, but the amounts consumed is quite small compared to Nd.[19] Although an important market, the amount of Nd consumed is small compared to the amounts of other rare earths that are used as phosphors.

Another important optical use is that of $La_2O_3$ in camera lenses and other optical devices. The $La_2O_3$ is added to fine optical glass to increase the index of refraction, while lowering the dispersion of the light passing through the glass. In this application the La purity can be of the order of 99.99%.

The final optical application I wish to discuss is the the electro-optic material called PLZT, lanthanum (L) modified lead (P) zirconate (Z) - lead titanate (T).[21] This material is a ferroelectric ceramic which is, under certain conditions, transparent to light and at other times opaque. PLZT is used as shutters and modulators, thermal/flash protective and eye safety viewing devices, and image storage devices. The La purity requirement for this material is 99.9%, the lowest for the various optical applications involving rare earth materials.

## Magnetic Applications

Two of the most important permanent magnets involve rare earth metals, namely the $Nd_2Fe_{14}B$ and the $SmCo_5$ - $Sm_2Co_{17}$ base materials.[22] The latter were discovered in the mid-1960's and the former about 10 years ago. Both materials are strong permanent magnets with magnetic strengths, as measured in terms of an "energy product", much greater than those of the more common alinco and ferrite permanent magnets. The Nd-Fe-B is slightly stronger than the Sm-Co. The former is the largest market for an individual rare earth metal because of its high energy product, and because it uses cheaper starting materials which are available from politically reliable countries, which is not true for Co. The Sm-Co alloys are still important because they can be operated at much higher temperatures than the Nd-Fe-B magnets, which have an upper operating temperature of ~100°C.

The yearly growth of the Nd-Fe-B market is >20% and it is expected to continue to grow at this rate throughout the 1990's. One of the biggest markets for permanent magnets is the automobile, where magnets are used in the various motors (starting, window, fan, fuel pump, door lock, etc.), indicators and sensors, controls, speakers etc. Other important markets are voice coil motors for computer disk drives, magnetic resonance imaging (medical NMR scanners), motors and generators, speakers and microphones.

For the permanent magnet application 95 to 99% pure rare earth metals are sufficient. A percent or two of another rare earth element will not have much affect on the magnetic properties, and non-rare earth impurities are desirable to help pin domain wall motion and improve the permanent magnet properties.

Another interesting magnetic material is Terfenol - a ternary intermetallic compound, $(Tb_{0.3}Dy_{0.7})Fe_2$, which exhibits giant magnetostrictions in an applied field, 100 times those observed in Ni.[22] When a magnetic field is applied to a magnetostrictive material, it will expand or contract. Conversely, when stress is applied to the material, a magnetic pulse is generated. Some of the uses of terfenol include sonar devices, micropositioners and liquid control values. Again the purity requirements of the Tb and Dy is not stringent, 98 to 99% is sufficient. This market is quite small but it is expected to grow rapidly in the 1990's.

The last magnetic application I wish to discuss is the use of gadolinium gallium garnet (GGG) as bubble devices for domain memory storage.[1] In this application information is stored in the form of magnetic bubbles (polarities) on the tips of "tees" and "bars" of permalloy (~1/20 of the diameter of a human hair). The permalloy tees and bars are deposited on a thin film of a complex magnetic rare earth-iron garnet which has been epitaxially grown on a GGG single crystal substrate. The rare earth oxides used to prepare the garnets are about 99.9% pure.

## Magneto-optic Applications

Magneto-optic discs (the spelling used in optical technologies) for storage of information using amorphous lanthanide (Gd and Tb) - transition metal (Fe and Co) alloys became a commercial reality in the late 1980's. The amorphous alloy, magneto-optic discs can store 15 to 50 times more information (bit densities of ~$10^8/cm^2$ have been achieved) than the conventional magnetic hard disk (the spelling used in magnetic technologies). The other major advantage is that the near contact head to disk spacings are not required in the magneto-optic devices. The use of highly focused laser beams allows for the high storage density.

Films of $Tb_{25}(Fe_{0.9}Co_{0.1})_{75}$ 500 to 2000Å thick are prepared by either RF or DC sputtering. This film is coated by a transparent thin film ceramic, such as AlN, $Y_2O_3$ or $Si_3N_4$ to protect the magneto-optic film from oxidation and/or corrosion and also abrasion. The next generation material will contain Nd, replacing some of the Tb, because the Nd alloy has a higher Kerr rotation at long wave lengths (~400nm). One major problem is that the sputtering targets contain too much O from the lanthanide metals. Better lanthanide metals (i.e. those with lower O contents) would greatly aid in expanding this market.

This could be a sizable and growing market for the lanthanide metals, especially if the EDRAW (erasable direct read after write) discs are used in the home entertainment market, such as compact discs (CD) and to replace magnetic tapes currently used in cassettes and video cassette recorders (VCR).

Since the commercially available rare earth metals have purities which range from 95 to 98.at% pure with respect to all elements, this seems to be sufficient for this application, except as noted above that the O content could be lower. It is the light, interstitial elements (H, C, N and O) which are the major impurities in the metals on an atomic basis, and it is difficult to keep the concentrations of these elements low because the metals are very reactive, especially with respect to these four light elements and other electronegative elements, such as S, Si etc.

## Electrical and Electronic Applications

One of the oldest uses of the rare earths in the electrical and electronic area is the use of $LaB_6$ as an electron emitter. This application is approximately 40 years old. $LaB_6$ has a fairly simple structure, a CsCl-like structure with the $B_6$ atoms forming an octahedron which occupies the corner positions of the CsCl unit cell and the La atom the body-center position. This compound has a melting point in excess of 2500°C, an extremely low vapor pressure, a low work function, a high thermionic emission current and is a good electrical and thermal conductor. It is a better electron emitter than W and it is used in apparati, instruments and devices which need a bright source of electrons, such as electron microscopes, and plasma generators.[22]

Another application, which is nearly as old as the $LaB_6$ electron gun, is the use of rare earths, especially YIG (yttrium iron garnet), to control high frequencies (from the VHF radio frequencies to visible light) in microwave devices. For example, a YIG sphere is resonant over a decade and can be tuned by varying the strength of an applied external magnetic field. This property (behavior) accounts for the use of YIG as tuned oscillators, discriminators, preselectors and frequency multipliers. It is also used as a circulator in radar transmission systems, whereby it transmits RF energy in one direction of rotation but not the other. YIG's are also used to protect RF power tubes from sudden changes in the transmission line impedances and as antennas for receiving and transmitting microwaves.

A third long standing use of the rare earths in the electronics field is in ceramic barium titanate capacitors. The rare earth oxides, usually $Nd_2O_3$, are added to reduce aging, increase the capitance and especially to provide zero (and sometimes negative) temperature coefficients of capacitance. The rare earth additions vary from 0.5 to 50% by weight. These ceramic capacitors are based on paper-thin layers of the rare earth-doped barium titanate, and have been miniaturized for hybrid microcircuits.

The newest and probably the largest market for the rare earths, especially Y, in the electrical/electronic area is the use of yttria-stabilized zirconia (YSZ) as an O sensor and solid electrolyte. Yttria is added to stabilize the cubic fluorite structure of $ZrO_2$, and prevent or reduce the destructive monoclinic - tetragonal phase transition encountered in pure $ZrO_2$ during thermal cycling. In addition the $Y_2O_3$ increases the O ion conductivity, because when two Y atoms are substituted for Zr in the $ZrO_2$ a vacancy is created on the anion ($O^=$) sites. This defect structure leads to a high electrical conductivity which is essentially due entirely to O ion transport. At high temperatures YSZ is sensitive to the O concentration around it. When it is exposed on different sides to different O partial pressures a voltage is developed, which is proportional to these partial pressures as given by the Nernst equation. The most important use of the O sensors is in the automotive industry where YSZ electrodes are used to measure the O partial pressure in the exhaust gas from an engine. By using a feedback mechanism with the appropriate control circuitry the air to fuel ratio is maintained at the ideal stoichiometry to keep the exhaust emissions pollutants (unburnt hydrocarbons, CO and nitrous oxides) to a minimum.[23]

Other uses involve $(La,M)CrO_3$ as ceramic conductors, where M is usually a divalent metal, and rare earth-doped barium titanates as temperature-sensitive resistors (thermistors). In the former, the chromites are used as resistance heating elements especially in oxidizing atmospheres, and as electrodes.

In the latter application the addition of lanthanide to barium titanate lowers the high resistivity of the undoped titanate (~$10^{11}$ Ω cm) to ~$10^2$ Ω cm below the tetragonal/cubic transformation. Above this transformation temperature the resistivity rises extremely sharply, up to 25% per °C. Thus the lanthanide barium titanate can be used as a heat-activated switch. The lanthanide (La, Ce, Pr or Nd) concentration varies from 0.001 to 0.006 mol%, and by varying the concentration and the lanthanide ion added the tetragonal/cubic transformation temperature can be adjusted to any value between 60 and 180°C. These thermistors are used as a safety cut-off switch in electric motors if the windings get too hot.

In all the applications outlined above in the electrical and electronics area, the purity requirements are of the order of 99% pure. In general, concentrations of non-rare earth elements need to be more stringently controlled than those of the rare earth element impurities.

## Future Applications

Research developments reported over the last few years suggest that there will be many new and exciting applications in the electronics field involving rare earth materials. Some of these include: the high temperature $YBa_2Cu_3O_{7-x}$ superconductor, $LaF_3$ thin film O sensors; the $RFe_{12-x}M_x$ family of high strength permanent magnets; $LaNi_5H_x$ batteries; $ErSi_2$ low resistance contacts and connectors in VLSI (very large scale integrated) circuits; the PbEuSeTe/PbTe mid-infrared laser diodes; and heavy metal fluoride glass optical fibers.

Although several other higher transition temperature superconductors have been found since the discovery of 90 K transition temperature in $YBa_2Cu_3O_{7-x}$ (1:2:3), the 1:2:3 phase will still be in the running for many of the high $T_c$ superconductor applications, because the two next important criteria (after having a $T_c > 77$ K) are the critical current capacity, $J_c$, and the fabricability into a useful form. In this regard the 1:2:3 is still the front runner, but there are still many hurdles which must be overcome. Commercial utilization of the high $T_c$ ceramics in simple devices is about ten years away and advanced applications such as supercomputers, levitated trains, electric generators and motors, magnetic resonance imaging units, and power transmission cables are at least 20 years away.

Another promising market for high purity rare earths (~99.999% pure) is in thin film electroluminescent display panels. In this application the emission of colored light from the phosphor is excited by an electric field rather than by electron radiation as in a CRT. In addition to using some of the rare earth phosphors (e.g. $ZnS:TbF_3$ for the green color, $Y_2O_2S:Eu$ for red and $SrS:Ce$ for blue), $Y_2O_3$ and $Sm_2O_3$ are being considered as dielectric layers between the electrodes and the phosphors. By adapting these materials to existing microelectronic techniques, thin multicolor display panels, with each pixel independently addressable via an x-y grid of electrodes, may be a reality within the next five to ten years.

The use of $LaNi_5H_x$ in rechargable batteries is quite likely to occur, because it will replace the current Ni-Cd rechargable batteries. The driving force will not be economics per se, but environmental concerns about the toxic nature of Cd in a device which is in common use throughout society today and will continue to be in the near foreseeable future.

The heavy metal fluoride glass optical fibers are receiving a great deal of attention because they have a predicted attenuation coefficient about an order of magnitude lower than silica glass optical fibers in use today. One of the major compositions being studied is call ZBLAN (ZBLAN has the composition, in mol%, 53 $ZrF_4$ - 20 $BaF_2$ - 4 $LaF_3$ - 3$AlF_3$ - 20 NaF). If the theoretical attenuation coefficient is realized, ZBLAN could be used to transmit information over a distance of 1500 km without a repeater. Furthermore, it has a second advantage over the silica glass optical fiber, in that ZBLAN is transparent in the IR region. This application will present the chemist and chemical engineers a real challenge since the impurity level requirements will be in the parts per billion range, about three orders of magnitude lower than is presently required for rare earth phosphors. The main problem will be to reduce the concentrations of the lanthanides which have unpaired 4$\underline{f}$ electrons down to the 1 ppb level, but other non-rare earth impurities, such as Fe could also present a problem.

**ACKNOWLEDGEMENT**

The author wishes to express his appreciation to J. Capellen and E. J. Calhoun for their critical comments and suggestions concerning this overview.

**REFERENCES**

1. K.A. Gschneidner, Jr., p. 403 in 'Speciality Inorganic Chemicals', R. Thompson, ed., The Royal Soc. Chem., London, 1981.
2. B.Z. Zhang, K.Y. Lu, K.C. King, W.C. Wei, and W.C. Wang, Hydromet., 1982, $\underline{9}$, 205.
3. D. Li, J. Chinese Rare Earth Soc., 1983, $\underline{1}$, [2], 4.
4. T.F. Anstett, "Availability of Rare-Earth, Yttrium, and Related Thorium Oxides - Market Economy Countries. A Minerals Availability Appraisal", U.S. Bur. Mines Info. Circular 9111, 1986.
5. J.G. Cannon, p. 1886 in 'Van Nostrand's Scientific Encyclopedia', 5th Ed. D.M. Considine, ed., Van Nostrand Reinhold Co., New York, 1976.
6. J. Kaczmarek, p. 135 in 'Industrial Applications of Rare Earth Elements', K.A. Gschneidner Jr., ed., ACS Symposium Series No. 164, Amer. Chem Soc., Washington, 1981.
7. R.J. Callow, 'The Industrial Chemistry of the Lanthanons, Yttrium, Thorium and Uranium', Pergamon Press, London, 1967.
8. J.E. Powell, p. 81 in 'Handbook on the Physics and Chemistry of Rare Earths', Vol. 2, K.A. Gschneidner and L. Eyring, eds., North-Holland, Amsterdam, 1979.
9. E.D. DeKalb and V.A. Fassel, p. 405 in 'Handbook on the Physics and Chemistry of Rare Earths', Vol.

4, K. A. Gschneidner, Jr. and L. Eyring, eds., North-Holland, Amsterdam, 1979.
10. J.W. O'Laughlin, p. 341 in 'Handbook on the Physics and Chemistry of Rare Earths', Vol. 4, K. A. Gschneidner, Jr. and L. Eyring, eds., North-Holland, Amsterdam, 1979.
11. R.J. Conzemius, p. 377 in 'Handbook on the Physics and Chemistry of Rare Earths', Vol. 4, K. A. Gschneidner, Jr. and L. Eyring, eds., North-Holland, Amsterdam, 1979.
12. A.P. D'Silva and V.A. Fassel, p. 441 in 'Handbook on the Physics and Chemistry of Rare Earths', Vol. 4, K. A. Gschneidner, Jr. and L. Eyring, eds., North-Holland, Amsterdam, 1979.
13. W.V. Boynton, p. 457 in 'Handbook on the Physics and Chemistry of Rare Earths', Vol. 4, K. A. Gschneidner, Jr. and L. Eyring, eds., North-Holland, Amsterdam, 1979.
14. R. S. Houk, in 'Handbook on the Physics and Chemistry of Rare Earths', Vol. 13, K. A. Gschneidner, Jr. and L. Eyring, eds., North-Holland, Amsterdam, to be published 1990.
15. R.J. Conzemius, S.-K. Zhao, R.S. Houk and H.J. Svec, Intern. J. Mass. Spectros. Ion Processes, 1984, 61, 277.
16. J.R. McColl and F.C. Palilla, p. 177 in 'Industrial Applications of Rare Earth Elements', K. A. Gschneidner, Jr., ed., ACS Symposium Series No. 164, Amer. Chem. Soc., Washington, 1981.
17. W. A. Thornton, p. 195 in 'Industrial Applications of Rare Earth Elements', K. A. Gschneidner, Jr., ed., ACS Symposium Series No. 164, Amer. Chem. Soc., Washington, 1981.
18. J. G. Rabatin, p. 203 in 'Industrial Applications of Rare Earth Elements', K. A. Gschneidner, Jr., ed., ACS Symposium Series No. 164, Amer. Chem. Soc., Washington, 1981.
19. M.J. Weber, p. 275 in 'Handbook on the Physics and Chemistry of Rare Earths', Vol. 4, K. A. Gschneidner, Jr. and L. Eyring, eds., North-Holland, Amsterdam, 1979.
20. L.W. Riker, p. 81 in 'Industrial Applications of

Rare Earth Elements', K. A. Gschneidner, Jr., ed., ACS Symposium Series No. 164, Amer. Chem. Soc., Washington, 1981.
21. G. H. Haertling, p. 265 in 'Industrial Applications of Rare Earth Elements', K. A. Gschneidner, Jr., ed., ACS Symposium Series No. 164, Amer. Chem. Soc., Washington, 1981.
22. K.A. Gschneidner, Jr. and A.H. Daane, p. 409 in 'Handbook on the Physics and Chemistry of Rare Earths', Vol. 11, K. A. Gschneidner, Jr. and L. Eyring, eds., North-Holland, Amsterdam, 1988.
23. F. L. Kennard III, p. 251 'Industrial Applications of Rare Earth Elements', K. A. Gschneidner, Jr., ed., ACS Symposium Series No. 164, Amer. Chem. Soc., Washington, 1981.

## Flat Panel Displays and Electronic Printing

High Information Content Display Trends in the 1990s
F. Funada (Sharp Corporation, Japan)

The Chemistry of Displays for the 1990s
I. C. Sage (BDH Limited)

Physics of Displays for the 1990s
E. P. Raynes (RSRE)

Dye Diffusion Thermal Transfer Printing (D2T2)
R. A. Hann (ICI Imagedata)

# High Information Content Display Trends in the 1990s

## F. Funada

OPTICAL DEVICE LABORATORIES, SHARP CORPORATION, TENRI, NARA 632, JAPAN

### 1 INTRODUCTION

The discovery of an electron beam by J.J.Thomson in 1897 has made the 20th century an electronics age accompanied with the early great inventions of a cathode ray tube(CRT) by Braun (1897) and a triode tube by de Forest (1906).

In addition, the superior inventions of a transistor by Schockley et al. and an integrated circuit (IC) by Kilby (1959) have also been making the latter half of this century an age of solid state electronics , where in particular the microelectronics which is today represented by VLSI has been constructing and establishing, what we call, a highly advanced Information Society.

In the society, there are several important elemental technologies; such as In-put terminals, Transmission, Memories, Processing and Out-put terminals. Their relationship is schematically shown in Fig.1.

At the same time, the examples of devices belong to each elemental technology are listed in Table 1.

From Fig.1 and Table 1, we can easily understand that the display device is located in the out-put terminals which acts as a man-machine interface and we understand that the display device gives a human enough information for his emotion or next decision-making.

Because of the recent support from a development of the electronics industry, the advanced Information-Society is abruptly going to grow up more and more, consequently the display is strongly required very high degree of performance from the consumer and markets.

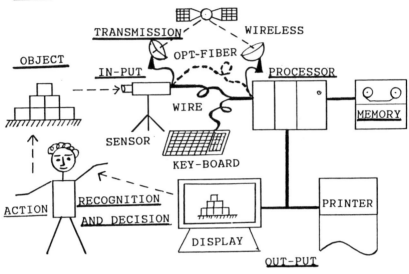

Fig. 1  The Information-Society

Table 1    Devices in the Information-Society

| FUNCTION | EXAMPLE OF DEVICES |
|---|---|
| IN-PUT | ♦ KEY BOARD, MOUSE, HANDWRITING, etc<br>♦ SENSOR (TRANSDUCER) |
| TRANSMISSION | ♦ WIRE<br>♦ WIRELESS (SATELLITE)<br>♦ OPTICAL FIBER<br>♦ MEDIA (MAG-TAPE, VD, OPT-DISC etc) |
| PROCESSOR | ♦ COMPUTER<br>   SUPER C., LARGE C., MINI C.,<br>   PERSONAL C., MICRO C. etc |
| MEMORY<br>(MEDIA) | ♦ SEMICONDUCTOR MEMORY (DRAM, SRAM, ROM)<br>♦ MAGNETO-OPTICAL DISC<br>♦ VIDEO DISC<br>♦ MAGNETIC TAPE |
| OUT-PUT | ♦ DISPLAY (DIRECT VIEW, PROJECTION)<br>♦ HARD COPY (DOCUMENT, PICTURE) |

Their requirements are,
- high resolution
- large picture size
- high luminance
- high contrast
- wide color gamut (grayscale)
- wide viewing angle
- fast response speed
- thin thickness
- light weight
- low power consumption
- high reliability

and
- low cost

These items are briefly summarised as the characteristics of high information content, good legibility and easy operation.

In this lecture, the high information content display trends in the 1990's are discussed from the points of application markets and suporting technologies.

2  Market trends in the 1990's

When the trends of display technologies are discussed, it is very effective to parallelly study the trends of the other related technological fields.

For example, let us consider the development of ICs as memory and processing devices.

In the IC industry, a famous empirical evolution rule is well known. That is the Moore's law which indicates twice content growth per year with an optimistic prediction of future IC industry.

Fig.2 shows the relationship between the number of transistor elements in one DRAM chip and its developed year, which verifies the Moore's law with good accuracy. Even though the slope of the Moore's law have been getting low (about twice per two years) after 1980, the growth tendency still continues with a strong activity. Therefore, the 16~64MDRAM has been just researched and the 1GDRAM is expected to be developed by the end of this century.

Not only in the DRAM but also in the other technologies; sensors, optical memories, transmission and in-put/out-put terminals, we can find similar situation and tendency.

From this background, the display devices as an especially important out-put devices have been strongly desired to improve in the characteristics which are particularly related to the information content for full and perfect translation of the vigorous processed information.

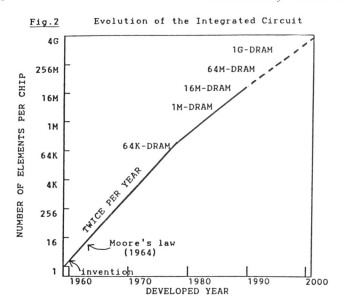

Fig.2 Evolution of the Integrated Circuit

Table 2  Displays and their principles

| DISPLAY | PRINCIPLE |
|---------|-----------|
| CRT | Cathodeluminescence by a deflected high speed electron beam (~20keV). |
| Flat-CRT | Similar to a conventional CRT only configurated in a flat format. May or May not use deflection. |
| VFD | Cathodeluminescence by a slow speed electron beam (~100eV). |
| PDP | Emission from a low pressure gas discharge. |
| LED | Emissin from a carrier recombination radiation with an electron and a hole. |
| EL | Emissin from a collision ionization by an accerarated electron. |
| LCD | Electro-optical effects caused by molecular alignment deformation of the LC. |
| ECD | Electrochromism from an electro-chemical reaction. |
| EPD | Electrophoresis of a colored dispersed particle. |
| SPD | A electrically (or magnetic) reorientation effect of colored suspended particles. |
| ESD | Electro-Scopic Display |
| OTHERS | Electret Displays, Elastomer Displays Magneto-optic Displays. |

Consequently, many kinds of displays based on the different principles have been researched.

Typical examples are listed in Table 2.
Among them, we can pick up several displays as the high information content displays.
They are CRT(Cathode Ray Tube), F(Flat)-CRT, LCD(Liquid Crystal Display), PDP(Plasma Display Panel), EL(Electroluminescence), and VFD(Vacuum Fluorescent Display).
Table 3 shows recent development status of the major candidate technologies for the high information content displays.[1~12]

By the way, let us turn preview the display market trends in the 1990's.[13]
It is no doubt that the biggest display markets are existed in both computers and televisions.

Remarkable advance of the VLSI have made the computers not small size but easy to use and low cost so much, hence the computers will have been getting very friendly personal stationery like a pen or a notebook in near future.
Last year(1989), about 20 million sets of the personal computers were produced about in the world, and in 1995, their market will grow up about twice scale, 40 million sets per year, and 60% out of them, 24 million sets must be portable types.
Accordingly, even if we limit to the portable computers, sales amount of flat panel displays is predicted more than 7 billion $.

Fig.3 shows the world-wide display market trends in the first half of 1990's, which include total application fields. The major parts are expected to be shared by the portable personal computers and the high definition television(HDTV).

If this prediction is correct, the total production amount of flat panel displays will overcome that of the CRT in 1996.
In particular, if the LCD will have been growing with the same rate, the king-position of the displays will be taken by the LCD at the end of this century.[14] At the time the CRT will be just celebrating its centenary.
Thereby, many Japanese electronics companies have decided to make a large amount of investment (total 700 million $ per year) on the high information content LCD in this year,1990.

Another promising market is the HDTV.
The HDTV can be thought as a super set of advanced television, it would have about twice the horizontal and vertical resolution of today's TVs, a wide aspect ratio of 16 x 9 for a spectacular scene and a better digital audio quality.

Table 3    Recent High Information Content Displays

| Technology | Picture-Size (diagonal V") | Number of Pixels | Full-Color | Brightness (cd/m2) | Others |
|---|---|---|---|---|---|
| CRT (Direct V.) | 43 | 2265 x 560 | YES | 110 | 115kg,35kV,71cm TV, SONY[1] |
| CRT (Direct V.) | 28 | 2048 x 2048 | YES | 70 | 98kg, D=671mm, CAD/CAM,SONY[2] |
| CRT (Proj.) | 200 | 1700 x 1000 x 3 (RGB) | YES | 33 (SG=2) | 250kg,34kV HDTV,Mithubishi[3] |
| Flat-CRT | ⟨40⟩ 12 | ⟨4524 x 1044⟩ 364 x 440 | YES | 230 | MDS multiunits HDTV,Matsushita[4] |
| LCD (Direct V.) | 14 | 1920 x 480 | Multi(512) | 100 | DSTN-LCD,28V 35mmD, SHARP[5] |
| LCD (Direct V.) | 14 | 642 x 480 (1284 x 960)sub-px. | YES | 120 | a-Si TFT-LCD,[6] 27mmD, SHARP |
| LCD (Proj.) | 240 | 1250 x 1250 x 3 | YES | 50 | Photocond.-add. 150kg, Hughes[7] |
| LCD (Proj.) | 110 | 1200 x 1000 x 3 | YES | 160 | a-Si TFT-LCD [8] HDTV,   SHARP |
| LCD (Proj.) | 38 | 2000 x 2000 x 3 | Multi(8) | 130 | Ferroelect.LCD CAD,Matsushita[9] |
| PDP | 33 | 1024 x 800 | YES | 58 | DC-PDP,0.11m/W ~250V,NHK(JBC)[10] |
| VFD | 8.3 | 320 x 200 | YES | 35 | 150V, 65mmD Futaba,[11] |
| EL | 6 | 960 x 240 | YES | 2.3 | Hybrid stack.EL ~200V, Planer [12] |

### Fig. 3 Display Market Trends

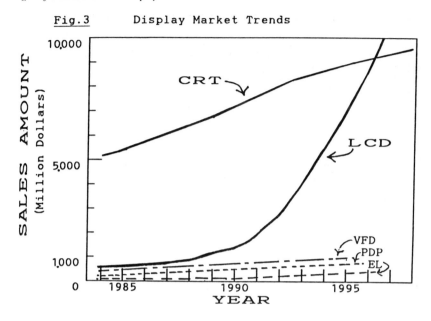

### Table 4 Displays for the Personal Computer & the HDTV

| | ITEMS | | COMPATIBILITY | | | | | |
|---|---|---|---|---|---|---|---|---|
| | | | LCD | CRT | F.CRT | PDP | EL | VFD |
| PERSONAL COMPUTER (book type) | PICTURE SIZE | ~10" | ⦿ | ⦿ | ⦿ | ⦿ | ⦿ | ◊ |
| | PIXEL | 640x480x3 | ⦿ | ⦿ | ⦿ | ⦿ | ⦿ | X |
| | COLOR | FULL/MULTI | ⦿ | ⦿ | ⦿ | ◊ | ◊ | ⦿ |
| | BRIGHTNESS | 〉100nt | ⦿ | ⦿ | ⦿ | X | X | ⦿ |
| | DEPTH | 〈20mm | ⦿ | X | X | ⦿ | ⦿ | ⦿ |
| | POWER | 〈10W | ⦿ | X | X | X | X | X |
| | WEIGHT | 〈2kg | ⦿ | X | X | ⦿ | ⦿ | X |
| | RESPONSE | 〈30msec | ⦿ | ⦿ | ⦿ | ⦿ | ⦿ | ⦿ |
| | availability | | ⦿ | X | X | X | X | X |
| HDTV (home use) | PICTURE SIZE | 〉50" | proj. ⦿ | proj. ⦿ | ◊ | ⦿ | ◊ | X |
| | PIXEL | 1500x1000x3 | ⦿ | ⦿ | ⦿ | ◊ | ◊ | X |
| | BRIGHTNESS | 〉300nt | ⦿ | ⦿ | ⦿ | ◊ | X | ◊ |
| | FULL COLOR | | ⦿ | ⦿ | ⦿ | ◊ | ◊ | ◊ |
| | RESPONSE | 〈30msec | ⦿ | ⦿ | ⦿ | ⦿ | ⦿ | ⦿ |
| | WEIGHT | 〈50kg | ⦿ | ◊ | ◊ | ◊ | ◊ | X |
| | availability | | ⦿ | ⦿ | ◊ | ◊ | X | X |

⦿:good, ◊:fair/possible, X:poor

Industrial importance of the HDTV is not only realizing a beautiful picture but also creating the novel application fields of the video tools, for example, there are VTR/VD movie theaters, printing/publishing, education, medical equipments, museums and so on......
As a result, the total market size of the HDTV is estimated 7 billion $ in 1995 and 35 billion $ in 2000,[15] in which displays would share about one third of them.
Also in the HDTV, the display is one of the important key technologies for its industrial success the same as a semiconductor.
Obviously, displays for the HDTV must essentially possess a large picture size, a high resolution and a good color gamut from its concept, therefore the LCD is expected as the most promising one.

Table 4 summarizes specifications of the portable personal computers and the HDTVs as well as the evaluation of applicability for the major candidative displays.

### 3 LCD as the first candidate

Recently, it is said that the quality of the LCDs have been improved very rapidly. In particular, the information content has been increasing remarkably.
When the LCDs introduced first into the market with massproduction in 1973, the LCDs had only about 30 to 70 pixels for the application of wrist watches and handy calculators.
Of course, they could perform neither half-tone nor color displays.
Today, after 17 years, the LCDs have grown up to be recognized as the powerful candidate for the high information content displays.
In this section, the trends of the LCD are reviewed from the points of materials and device structures.

<u>Liquid Crystal Materials</u>

The first massproduced LCD[16] used in the DSM[17] (Dynamic Scattering Mode) is a mixture of the Schiff's base compounds, MBBA[18] and its homologues, and the quaternary ammonium salt as an ionic dopant.[16]
However, by 1975 the DSM-LCD was replaced by the TN(Twisted Nematic)-LCD.[19] Because the DSM demanded relatively high driving voltage (about 8 v), hence it had not good compatibility with the CMOS-LSI drivers.
The Schiff's base compounds with a terminal -CN group[20] was used for the early TN-LCDs for the positive

dielectric anisotropy, so that the LCDs suffered from a fatal blow in a reliability problem owing to the hydrolysis.

The cyano biphenyl compounds emerged as a salvation, which was developed by Gray et al. in 1973.[21] They were the first materials which had good chemical stability, non optical absorbtion within the visible light, practical nematic phase range and good electrooptical characteristics, therefore this material should be memorable in the history of the LCD the same as the first room temperature nematic compound, MBBA.

Since the success of the biphenyl compound, the concept of direct linkage of core rings has become standard tactics.

In order to improve the response and viewing angle of the TN-LCDs, both lower viscosity and small birefringence had been required strongly.

As the result, cyclohexane ring compounds (ECH) were proposed by Demus et al.[22]

Following to them, many kinds of novel useful mesophase compounds have been synthesized such as; PCH[23], CCH[24], PYP[25], PDX[26], EPCH[27], Tolan moiety[28], etc[29~31] and multi ring compounds like a CBC[32], which have made a great contribution to grow the today's LCD industries supporting with many severe physical requirements from the TN and the super TN(STN)-LCDs.

These typical features are summarized in Table-5.

By the way, in order to obtain the best performance of the LCDs, an active matrix addressing scheme[33] has introduced into the comsumer market in the middle of 1980's.[34]

In the active matrix addressed LCDs, the liquid crystal materials must possess not only an appropriate electrooptical properties but also a good dielectric function which stores adequate electric charge for the frame time interval.

Thereby, the liquid crystal materials for this application should have a very high resistivity (e.g. more than $10^{14}$ OHM cm, at $25°C$). From this point of view, instead of the conventional cyano-compounds recently developed fluorinated compounds[35]; with-F, -F.F, -CF3, -OCF3, are very interesting and expecting.

In addition to the nematic phase, the smectic A and chiral smectic C phases have been studying for the high information content displays by utilizing their inherent thermal[36] or ferroelectric effects[37] respectively.

The most currently used smectic A materials are alkyl(alkoxy) cyano-biphenyls.

Using the thermal effect of the smectic A phase, a laser

Table-5  Typical Liquid Crystal Compounds

| Abbreviation | Compound | Comment |
|---|---|---|
| PAA[31] | $CH_3$-⟨⟩-N=N(O)-⟨⟩-$OCH_3$ | C 117°C N 135°C I |
| APAPA[17] | $CH_3O$-⟨⟩-CH=N-⟨⟩-$OCCH_3$ (=O) | C 83°C N 110°C I |
| MBBA[18] | $CH_3O$-⟨⟩-CH=N-⟨⟩-$C_4H_9$ | C 22°C N 47°C I |
| N | $CH_3O$-⟨⟩-N=N(O)-⟨⟩-$C_2H_5$ | C 16°C N 76°C I |
| E [20] | $C_nH_{2n+1}$-⟨⟩-CO-O-⟨⟩-CN | large $\varepsilon_\parallel$, high $\eta$ |
| CB [21] | $C_nH_{2n+1}$-⟨⟩-⟨⟩-CN | stable, well balanced |
| ECH[22] | $C_nH_{2n+1}$-⟨H⟩-CO-O-⟨⟩-CN | low $\eta$, low $\Delta n$ |
| PCH[23] | $C_nH_{2n+1}$-⟨H⟩-⟨⟩-CN | do. |
| ECP[27] | $C_nH_{2n+1}$-⟨H⟩-$CH_2CH_2$-⟨⟩-CN | do. |
| PYP[25] | $C_nH_{2n+1}$-⟨N,N⟩-⟨⟩-CN | large $\Delta n$, low $K_3/K_1$ |
| T[28] | $C_nH_{2n+1}$-⟨⟩-C≡C-⟨⟩-$C_mH_{2m+1}$ | high $\Delta n$ |
| CBC[32] | $C_5H_{11}$-⟨H⟩-⟨⟩-⟨⟩-⟨H⟩-$C_3H_7$ | C 54°C $S_1$ 232°C $S_2$ 251°C N 312°C I |
| PCH·F[35] | $C_nH_{2n+1}$-⟨H⟩-⟨⟩-F | very low $\eta$ & $\Delta n$, high $\rho$ |
| PCH·$OCF_3$[35] | $C_nH_{2n+1}$-⟨H⟩-⟨⟩-$OCF_3$ | do. |
| DOBANBC[39] | $C_{10}H_{21}O$-⟨⟩-CH=N-⟨⟩-CH=CH-CO-O-$CH_2$-*CHC_2H_5$($CH_3$) | C 76°C $S_C$ 95°C $S_A$ 117°C I |
| γ-Lactone[41] | $C_5H_{11}$-⟨H⟩-⟨⟩-O*$CH_2$-*⟨lactone⟩-$C_4H_9$ | Optically active dopant |
| γ-Lactone[42] | $C_8H_{17}$-⟨N,N⟩-⟨⟩-O*$CH_2$-*⟨lactone⟩-$C_4H_9$ | High Ps (~600nC/cm²) |
| MHPOBC[44] | $C_8H_{17}$-⟨⟩-⟨⟩-CO-O-⟨⟩-CO-O-*$CH$-$C_6H_{13}$ ($CH_3$) | Antiferroelectric phase |

diode beam addressed very high content LCD[38] was developed by several companies, whose content had already exceeded the CRT's one.

Another approach is the application of the ferroelectricity of the chiral smectic C phase. In 1975 Meyer et al[39] predicted theoretically and verified the ferroelectricity experimentally based on the symmetrical consideration of the chiral smectic C phase
which has monoclinic point group C2 using the compound; DOBAMBC.
And in 1980 Clark and Lagerwall invented an ingenious ferroelectric LCD[37], they called it SSFLCD(Surface stabilised ferroelectric-LCD), using also this compound
By the half of 1980's, it was a general way to mix chiral smectic C compounds themselves for obtaining an experimental mixtures. However this way is not profitable because of their high viscosity and high cost due to their chiral branched group[40].
Therefore, today, the chiral smectic C phase is made of non-chiral smectic C hosts and chiral dopants[41][42].
Using this approach, the ferroelectric liquid crystal mixtures have been improved rapidly, then some
prototypes of the high information content LCD[9][43] have been already developed and demonstrated. However, their performance is not yet adequate for the practical usage especially in the contrast and the response time, hence a revolutionary compound is strongly desired.
More recently, antiferroelectric phase[44] has been discovered, and is expected to be applied to the high information content LCD because of the definite electrooptical threshold and tri-stable characteristics

Recent examples of optically active dopants[41][42] which induce a high spontaneous polarization and the antiferroelectric compounds[44] are also shown in Table-5.

## X-Y matrix LCD

The X-Y matrix addressing scheme is the most currently used method for the high information displays for decreasing the number of the drivers and connections drastically compared with the static driving. On the other hand, in this addressing scheme, only 1/N of frame time, where $N$[45] is a number of scanning line(Y line), is effective for the excitation, so that average driving energy for the selected pixels must decrease depending on N, and as a result, it emerges a poor display contrast.
In order to break through this contradiction, two different technical approaches have been studied.

One is improvement of the transfunction of the

electrooptic effects itself based on the liquid crystal materials and the LCD structures using the conventional simple X-Y matrix electrodes structure. Then, this approach is called as a simple matrix addressing scheme.

The other is to build electrically non-linear active elements, like diodes or transistors, in the X-Y matrix LCD, what we call an active matrix LCD[33].

Simple Matrix LCD

If we apply the conventional TN mode to the simple matrix LCD with a high multiplexing rate, a strict problem will emerge because of its inherent steepless transfunction even though many efforts[46~47] had been done for the improvements from the various aspects; liquid crystal materials, panel constructions and driving methods.

Since the end of 1970's, several groups[48~50] have found super twisted nematic (STN) configurations which bring remarkably steep electrooptic characteristics.[51]
One of the residual important problems of the STN LCD was an interference colored appearance.
Various attempts have been proposed for this problem, then the most successful one is the application of an optical compensator(phase retarder) like the Babinet compensator in the optics.
Representative example of this approach is the double layered STN(DSTN) LCD whose concept was invented in 1981 by us.[52]

Fig.4  A Structure of the DSTN-LCD

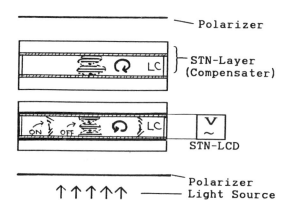

A schematic structure of the DSTN-LCD is shown in Fig.4, which is made of a conventional STN panel and a reversally twisted electrodeless STN layer which has the same retardation value and is setted at a crossed liquid crystal molecular orientational configuration.
The DSTN LCD has both steep and panchromatic electrooptic characteristics, hence we can realize a legible display. In addition to the B/W display, combining a RGB color filter we can also make a multi/full color display using the DSTN technology[5].[53]
Another approach using the compensation method have been studying actively utilizing plural polymer retardation films instead of the second STN layer for the purpose of deminishing the weight and the cost of the device, which is called the film compensation (F)STN LCD.[54~55]
Consequently, the performance of the FSTN LCDs have been suprisingly improved, but they have not yet achieved the level of the DSTN LCD in the display quality, hence we have been making many efforts of optimizing the optical properties and configurations of the films.
Another expected method in the simple matrix LCD is the SSFLCD[37] applying the chiral smectic C phase.
The SSFLCD has following outstanding features.
- ◆ no limitation of a number of the scanning line even for the simple matrix LCD because of its inherent memory effect and fast response.
- ◆ broad viewing angle
- ◆ panchromatic characteristics

Recently, three outstanding developments have been reported in this field.
The first one is the direct view type large size (14") high content (1280 x 1120) display for the EWS by CANON[43]. The second is the high content (2000 x 2000 x RGB) rear projection (38") display for the CAD application by Matsushita[9].
The third is the prototype display (32 x 64) using the antiferroelectric phase by Nippondenso[56].
However, in order to be competitor of the active matrix LCD and CRT, we need to resolve two residual problems for uniform orientation with bistability and fast response.

### Active Matrix LCD

In the active matrix LCD non-linear switching elements are integrated in each cross points of the X-Y matrix array.
Applying this addressing scheme, we have been clearly set free from the contradiction problems between the information content and the optical performance.
Because the driving voltage in the selected and non-selected states are not restricted.

The most practical switching active element is a thin film transistor (TFT) utilizing a-Si[57] or p-Si[34] as a semiconductor material.
The commonly used display mode of the full-color active matrix LCD is the TN mode for its relatively balanced electrooptical characteristics.
The TFT-array is fabricated from the thin film deposition processes like a CVD and a sputtering and the photolithography processes which are currently used in the Si-LSI technology.
Fig.5 shows a picture of 14"V, 642 x 480 x4 dots full color image display using the a-Si TFT active matrix LCD developed in 1988 by SHARP[6].

In addition to the direct view display, projection types using an active matrix LCD as a light valve have also been developed, for realizing a very large (more than 100" diagonal) and high resolution picture. This is also continuously applied to the HDTV.[8] [58]

The residual biggest future subject of the TFT-LCDs is just to cost issue. In particular, it requires to reduce cost of the key elemental parts such as the TFT-array, the micro-color filter, and the peripheral LSI drivers[6) 59~61].

Concerning to the TFT array, every effort has been devoted to prevent defects by utilizing various redundant design and laser-assisted amendment processes and, of course, to reduce the total fabrication processes for increasing the productivity.

On the micro-color filters, for simplifying the fabrication process, the improved photolithography[62] (pigment dispersed photo-polymers), the printing[63], and the electrodeposition methods[64] have been investigated

And on the drivers, the p-Si TFT active matrix LCDs have been taken a strong interest in realising the monolithically integrated peripheral driver circuit[64~65] on the same glass substrate utilizing its high mobility, which should bring not only broad cost-down but also high reliability and compactness.
A recent topics for this item was fundation of a research corporation, the Giant-electronics Technology Corporation (GTC) established in the last year (1989) under the direction of the Ministry of International Trade and Industry (MITI) of Japan, which aimed to develop a very large size direct-view full-color LCD with monolithic peripheral driving circuits using the p-Si TFT technology and was organised by 17 companies including two European relating companies.

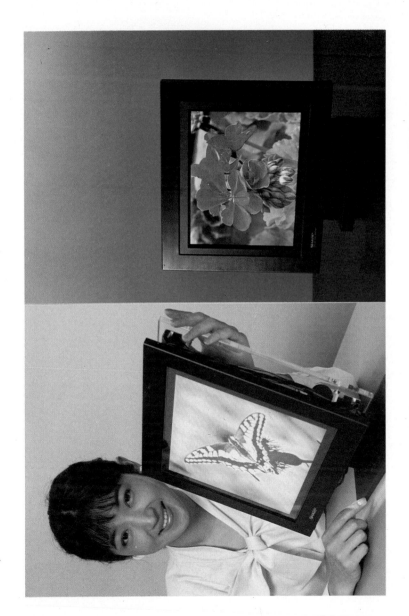

Fig.5  14"V (642 x 480) Full Color TFT-LCD

## 4 CONCLUDING REMARKS

In the world wide display industries, great evolution and revolution will emerge in the 1990's the same as the politics and the economics of the Europe.

The flat panel display is strongly expected to get the new sovereign replacing the traditional CRT in the biggest markets of the high information content display which are desired from the highly sophisticated Information Society based on the huge semiconductor and electronics industries.

In particular, the TFT active matix LCD is considered to be the most promising one, which could satisfy the severe qualitical requirements for the novel big markets of the portable personal computers and the HDTVs.

REFERENCES

1. M.Nakamura et al, SID 88 Digest, 1988, 386
2. H.Ohno et al, SID 89 Digest, 45
3. S.Shindo et al, Mitsubishi Tech. J., 1989, 63, 5
4. K.Nonomura et al, SID 89 Digest, 1989, 106
5. SHARP, Expo.at the Japan Electronics Show,Oct.1989
6. T.Nagayasu et al, Proc.Int.Disp.Res.Conf., 1988, 56
7. A.G.Ledebuhr, SID 86 Digest, 1986, 379
8. SHARP, Expo.at the Japan Electronics Show,Oct.1989
9. Y.Iwai et al, Proc.Japan Display 89, 1989, 180
10. H.Murakami et al, Proc. Japan Display 89, 1989,214
11. K.Morimoto et al, Proc. Japan Display 86, 1986,516
12. C.N.King, Proc. Eurodisplay 87, 1987, 14
13. Nikkei BP ed., Flat Panel Display (Jap.) Oct. 1989
14. W.E.Howard, Proc. Japan Display 89, 1989, 8
15. Data sheets issued by Hi-Vision Promotion Cent.1989
16. S.Mito & T.Wada, 5th Int.Liq.Cryst.Conf.1974,247
17. G.H.Heilmeier et al , Proc.IEEE, 1968,56, 1162
18. H.Kelker & B.Scheurle, Angew.Chem., 1969, 8, 884
19. M.Schadt & W.Helfrich, Appl.Phys.Lett.,1971,18,127
20. A.Boller et al, Proc. IEEE, 1972, 60, 1002

21. G.W.Gray et al, Electronics Lett., 1973, 22, 130
22. D.Demus, Nonemissive Electrooptic Disp.Plenum,1976
23. L.Pohl et al, Phys.Lett., 1977, 60A, 421
24. R.Eidenschink et al, Angew. Chem., 1978, 90, 133
25. A.Boller et al, Mol.Cryst.Liq.Cryst., 1977, 42, 215
26. D.Demus & H.Zaschke,Mol.Cryst.Liq.Cryst.1981,63,129
27. M.J.Bradshaw et al, Mol.Cryst.Liq.Cryst.1983,97,177
28. H.Takatsu et al, Mol.Cryst.Liq.Cryst. 1986,141,279
29. B.S.Scheuble, Proc. Euro-Display 87, 1987, 88
30. E.Poetsch, Kontakte(E.Merck), 1988,2,14
31. D.Demus, Mol.Cryst.Liq.Cryst. 1988, 2, 14
32. R.Eidenschink et al,Proc.Freiburg.Arb.Flus.K.1980,2
33. B.J.Lechner et al, Proc.IEEE, 1971, 59, 1566
34. S.Morozumi et al, SID 83 Digest, 1983, 156
35. H.A.Kurmeier et al, Int.Pat.Appl.WO89/02884,1987
36. H.Melchior et al, Appl.Phys.Lett., 1972, 21, 392
37. N.A.Clark & S.T.Lagerwall,Appl.Phys.Let.1980,36,899
38. A.G.Dewey & J.D.Crow, IBM J.Res.Dev., 1982,26,177
39. R.B.Meyer et al,J.de Phys.,1975, 36, L69
40. K.Skarp et al, Mol Cryst.Liq.Cryst, 1988, 165,437
41. M.Koden et al, Proc.Japan Display 89, 1989,34
42. M.Koden et al, Jap.Chem.Soc.Spring-Mtg.Abst,1990
43. H.Inoue et al, Int.Disp.Res.Conf.Post-D.paper,1988
44. A.D.Chandani et al,Jap.J.Appl.Phys, 1988, 27, L729
45. P.M.Alt & P.Pleshko, IEEE trans.ED,1974, ED21,146
46. G.Baur, Mol.Cryst.Liq.Cryst., 1981, 63, 45
47. M.Schadt, Mol.Cryst.Liq.Cryst., 1988, 165, 439
48. T.Hiroshima et al,Jap.Appl.Phys.Spr.-Mtg.30PB5,1979
49. C.M.Waters et al, Proc.Japan Display 83, 1983, 396
50. T.J.Scheffer, Proc.Japan Display 83, 1983, 400
51. T.J.Scheffer et al, Appl. Phys. Lett., 1984,45,1021
52. F.Funada et al, UK Pat. GB2092769B,1984
53. N.Kimura et al, SID 88 Digest, 1988, 49
54. M.Ohgawara et al, SID 89 Digest, 1989, 390
55. M.Akatsuka et al, Proc.Japan Display 89, 1989,336
56. M.Yamawaki et al, Proc.Japan Display 89, 1989, 26
57. W.E.Spear & P.G.LeComber,Sol.St.Commun.1975,17,1193
58. H.Noda et al, Proc.Japan Display 89, 1989, 256
59. H.Moriyama et al, SID 89 Digest, 1989, 144
60. H.Nakajima et al, SID 89 Digest, 1989, 234

61. Y.Matsueda et al, SID 89 Digest, 1989, 238
62. S.Okazaki et al, Jap.EID88-53(Jap.) 1988,1
63. T.Uchida et al, Proc. Euro-Display81, 1981, 39
64. M.Suginoya et al, Proc.Japan Display 83, 1983, 206
65. F.Emoto et al, Proc.Japan Display 89, 1989, 152

# The Chemistry of Displays for the 1990s

## I. C. Sage
ORGANIC DEVELOPMENT DEPARTMENT, BDH LTD., BROOM ROAD, POOLE, DORSET BH12 4NN, UK

### INTRODUCTION
The progress of chemistry for displays in the 1990's is a story that will take a full decade to unfold, and in an article such as this it is necessary to simplify the argument by making some assumptions. It is the author's contention that although the dominant display technology for applications requiring a moderately complex information format will continue to be the cathode ray tube, the principal area for *innovation* in the field will be in passive display technologies. Throughout the 1980's, the trend in display development has been to provide devices which offer a flat panel format, high information density, and improved legibility. Above all, however, has been a drive to increase the applicability of complex information displays by reducing their cost (on a per pixel basis) and making them suitable for inclusion in portable, industrial and in-car equipment etc. The need for displays to operate on battery power together with the cost benefit of using low power, low voltage CMOS driver circuits gives a decisive advantage to passive devices in many new fields of use.

The relevance of chemistry to the operation and manufacture of display devices is all-encompassing. Considering the cathode ray tube as an example, the range of chemistry involved can readily be appreciated:

| COMPONENT | CHEMISTRY |
|---|---|
| Glass envelope | Silicate chemistry |
| Screen | Phosphors |
|  | Rare earth chemistry |
| Shadow mask | Lithography |
| Electron gun | Metallurgy |
|  | Electron emissive materials |
| Insulators | Ceramics, Resins |

The author's background, however, means that this paper concentrates on a restricted range of organic chemistry applied to passive displays.

### PASSIVE DISPLAY TECHNOLOGIES
In recent years, the predominant developments in passive display technology have appeared

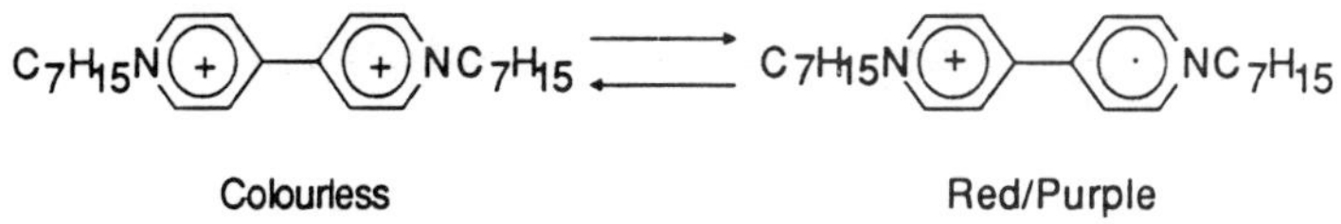

$C_7H_{15}N^+$—⟨+⟩—$NC_7H_{15}$ ⇌ $C_7H_{15}N^+$—⟨·⟩—$NC_7H_{15}$

       Colourless                              Red/Purple

Figure 1: Electrochromism in Di-heptyl viologen

in the field of liquid crystal (LC) devices. These displays now have a demonstrated capability to offer full colour rendering,[1] a resolution of more than 1000 lines,[2] and a directly viewed screen size of 20 inches diagonal.[3] All these qualities are combined with the proven performance of LC displays in respect of low operating power, matrix addressability and long operating life, and together represent a baseline which other devices must match if they are to find widespread acceptance in commercial applications. Although a number of alternative passive display effects have been the subject of development work over a long period, none yet appears to rival the performance of LC devices. Space does not allow a full discussion of all the device effects available; most of them have been hindered in their exploitation by problems of reliability, addressability or manufacturability. In order to illustrate the present limitations and progress in the development of alternatives to LC displays, a short account will be given of some recently published results on electrochromic devices.

Applied to a display device, unlike most LC effects, electrochromic materials require no polariser, and can provide an excellent viewing angle and good contrast and brightness. The resulting devices are usually single colour and rely on the passage of current for their operation. Electrochromism has been studied in a wide range of systems, including tungsten blues, lutetium phthalocyanine and its analogues, ferrocyanide complexes and conducting polymers. One of the most interesting systems, however, is the viologen radical-ion. Electrochemical reduction of solutions of eg. diheptylviologen leads to the deposition on the electrode of a purplish layer of the radical-ion (Figure 1). One of the effects which limits the lifetime of such a device is the tendency of the radical-ion to polymerise on the electrode to form a product which can no longer be oxidatively decoloured. The chemistry of inclusion complexes can be applied to this problem by adding cyclodextrins to the viologen solution. Long chain alkyl substituted viologen molecules readily associate with the hydrophobic cavity of the $\beta$-cyclodextrin cage; two moles of cyclodextrin appear to be required for optimum results. The complexed viologen radical-ion is now no longer able to approach its reactive neighbours sufficiently closely to form a polymer and a ten-fold improvement in operating lifetime has been reported for such devices.[4] A further benefit of the addition of cyclodextrin comes from a change in colour of the device. In the absence of complexation, the predominant coloured species is dimeric and reddish-purple in colour. The $\beta$-cyclodextrin cavity is only large enough to hold a single guest molecule and the resulting monomeric species has a more pleasant blue tone.

Although the above work does much to alleviate the limitation on write/erase cycles in electrochromic devices, several severe shortcomings remain to be adequately addressed, including the addressing technology for complex electrochromic displays and the provision

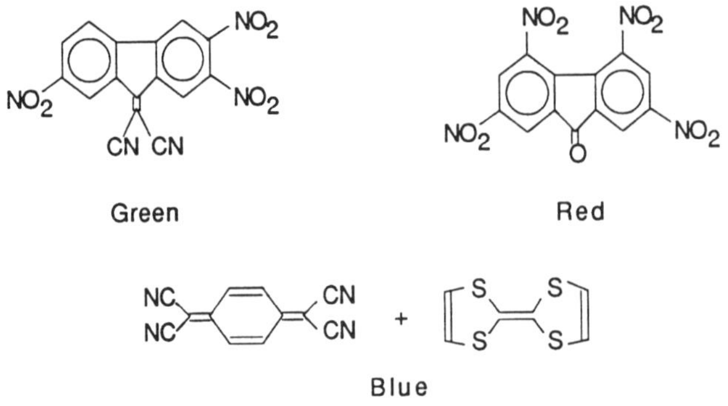

Figure 2: RGB Electrochromic Substrates

of colour. A full colour electrochromic display must rely either on a black on white display effect being overlayed on a passive colour matrix filter, or on electrochromic species of each primary colour being available. In the latter case, the species used must be insoluble to allow the use of a common electrolyte without migration of the active materials from one area of the electrode to another and progress in this direction has been reported.[5] The species shown in figure 2 are capable of providing electrochromic layers in the three primary colours when deposited as evaporated films on an electrode surface and are insoluble in an aqueous electrolyte. Unfortunately, the addressing time required to colour the films is as long as several seconds and the lifetime appears quite inadequate for practical use.

The above discussion is intended to illustrate both the shortcomings of electrochromic devices and the progress being made in their improvement. At the present time they do not appear to rival LC devices in their performance when applied to display applications but more decisive advances in this direction cannot be ruled out. Equally, there are non-display applications (such as controllable sunroofs for cars or dimmable rear-view mirrors) where electrochromic devices may hold the advantage. Similar comments might be made regarding other passive electro-optic effects such as electrophoretic or micromechanical devices. Pending significant advances in these areas however, LC displays seem likely to retain a dominant position in display development and the remainder of this article will concentrate on this theme.

## LIQUID CRYSTAL DISPLAYS
Liquid crystals have been exploited in a wide range of of electro-optic devices, each offering its own combination of qualities:

- Supertwisted nematic[6]
- Chiral smectic ferroelectric[7]

- Tunable birefringence[8]
- Twisted nematic[9]
- Polymer dispersed droplets[10]
- Electroclinic[11]
- Thermally addressed smectic[12]
- Dynamic scattering[13]

Of these effects, the first three have been used to demonstrate high complexity directly multiplexed displays while the twisted nematic device is dominant in simple format applications, but is also the effect most commonly used on an active matrix substrate to achieve very high performance devices. Each electro-optic effect requires a liquid crystal material optimised to obtain quite different physical properties for its successful operation, and the realisation of LC properties in desired combinations both by molecular engineering of new compounds and by formulating mixtures of different compound classes is now a mature science.[14] Examples of the properties which can be manipulated can readily be quoted:

| PROPERTY | STRUCTURAL DEPENDANCE |
|---|---|
| Clearing point | Core length |
| | Rigidity |
| | Electronic conjugation |
| | End groups |
| Birefringence | Electronic conjugation |
| Dielectric anisotropy | Electronic conjugation |
| | Dipolar substitution |
| | Control of dimerisation |
| Elastic constants | Core groups |
| | Lateral substitution |
| | Length of terminal chains |
| | Ethylene linking groups |

The design of novel LC materials is constrained by the need to provide colourless compounds of high stability and the variety of structural sub-units which are in common use is surprisingly limited. Table 1 lists some well established structures which are in common use, together with their most important physical properties.

Although a very large body of empirical knowledge has been accumulated which relates the physical properties of liquid crystalline compounds to their molecular structure, there is hardly ever a theoretical basis of sufficient generality and quantitative precision to give a useful guide to LC design. Taking as an example the most fundamental of liquid crystal properties, the LC to isotropic transition temperature or clearing point, attention has been drawn to the series of compounds in Table 2.[15] Starting from an alkyl cyanobiphenyl, an extraordinary divergence of behaviour is seen when one or other of the aromatic rings is saturated. Saturation of both rings again gives a highly stable LC phase, and analogous behaviour can be seen in the apparently similar 2-2-2-bicyclooctane derivatives. The design

| Structure | Clearing point | Dielectric Anisotropy | Birefringence |
|---|---|---|---|
| R–⟨⟩–⟨⟩–CN | 35 | 11.5 | 0.18 |
| R–⟨N⟩–⟨⟩–CN (pyrimidine) | 50 | 19.7 | 0.18 |
| R–⟨cyclohexyl⟩–⟨⟩–CN | 50 | 9.7 | 0.1 |
| R–⟨dioxane⟩–⟨⟩–CN | 50 | 13.3 | 0.09 |
| R–⟨cyclohexyl⟩–⟨cyclohexyl⟩–CN | 85 | 4.4 | 0.06 |
| R–⟨⟩–CO$_2$–⟨F⟩–CN | 30 | 48.9 | 0.14 |
| R–⟨⟩–CO$_2$–⟨⟩–R' | 20 | 0.5 | 0.13 |
| R–⟨cyclohexyl⟩–CO$_2$–⟨⟩–OR' | 75 | -1.0 | 0.07 |
| R–⟨cyclohexyl⟩–C$_2$H$_4$–⟨⟩–⟨F⟩–R' | 105 | 0.0 | 0.14 |

Table 1: Some Commonly Used Nematic Liquid Crystals

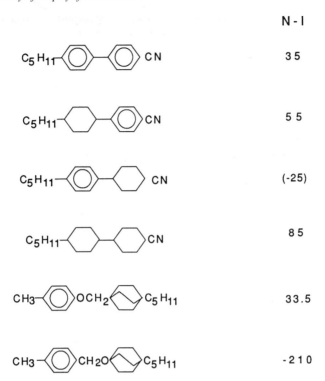

Table 2: N-I *vs* Structure in Nematic LC's

| LINES ADDRESSED (N) | VOLTAGE RATIO |
| --- | --- |
| 2 | 2.41 |
| 3 | 1.93 |
| 8 | 1.45 |
| 16 | 1.29 |
| 32 | 1.20 |
| 64 | 1.13 |
| 128 | 1.09 |
| 256 | 1.06 |
| 512 | 1.045 |

Table 3: Available Voltage Ratio in RMS Multiplexed Displays

'rule' in this case, that regions of high polarity or polarisability in a LC molecule must not be separated by regions of low polarisability, is well recognised and allowed for by chemists but still eludes an accessible theoretical explanation.

The amount of information which can be written onto a nematic LC display responding to the applied RMS voltage is limited by the fact that matrix addressing of such a device applies a definite voltage to areas of the display which should remain in the "off" state.[16] The ratio of voltage applied to "on" and "off" pixels is limited to:

$$V_{on}/V_{off} = ((N^{1/2} + 1)/(N^{1/2} - 1))^{1/2}$$

This ratio rapidly approaches unity as the number of lines of information on the display, $N$, increases. (Table 3). In order to generate good appearance in a display addressed in this way, an electro-optic effect is required which produces no visible change up to a certain applied voltage, then suddenly switches to produce its full contrast at a slightly higher voltage. This behaviour can indeed be realised in the supertwisted nematic (STN) device which has been rapidly exploited since its discovery in 1984.[6] The STN cell contains a nematic liquid crystal layer twisted through a total angle of 180–270 degrees,and behaves as a twisted optical retardation plate. Its manufacture is a further illustration of the central importance of chemistry in the displays area, relying as it does on the availability of special grades of glass, epoxy adhesives, polyimides to provide high-tilt alignment and polariser films as well as liquid crystal materials. In order to obtain the desirable switching characteristics referred to above, special requirements are placed on the properties of the liquid crystal, which may be contrasted with the needs of the twisted nematic display which pre-dated the STN effect in these applications.

C$_3$H$_7$-⟨H⟩-C$_2$H$_4$-⟨◯⟩-⟨◯⟩-C$_3$H$_7$     K 67 S 119 N 144 I

C$_3$H$_7$-⟨H⟩-C$_2$H$_4$-⟨◯⟩-⟨◯(F)⟩-C$_3$H$_7$     K 40 N 108 I

C$_3$H$_7$-⟨H⟩-C$_2$H$_4$-⟨◯(F)⟩-⟨◯⟩-C$_3$H$_7$     K 59 S (34) N 108 I

Table 4: The Effect of Lateral Fluorination on Smectic Phase Stability

| TWISTED NEMATIC | | SUPERTWISTED NEMATIC | |
|---|---|---|---|
| $K_{33}/K_{11}$ | Small | $K_{33}/K_{11}$ | Large |
| | | $K_{33}/K_{22}$ | Large |
| $\Delta\varepsilon/\varepsilon_\perp$ | Small | $\Delta\varepsilon/\varepsilon_\perp$ | Small |
| $K_{11}$ | Small | $K_{11}, K_{33}$ | Small |
| d$\Delta$n | 0.55, 1.05 | d$\Delta$n | 0.85–0.95 |
| | | High tilt alignment | |

It is some measure of the accumulated experience in mixture formulation, that satisfactory materials for *both* types of device can be made from the components in Table 1. The supertwisted nematic display can be driven by the RMS multiplex scheme referred to above, and generates adequate contrast in displays showing 500 lines of pixels. Improved supertwist displays use birefringent plastic sheets to compensate the optics of the panel and give black on white devices which are the displays of choice in the present generation of portable computers etc.

The synthesis of one of the materials[19] from Table 1 which is frequently used in STN devices is outlined in Figure 3 and illustrates some of the problems of preparation of LC's on a commercial scale.[20] Although each of the synthetic steps is in itself straightforward, the preparation totals some 14 steps in a poorly convergent sequence. As well as giving problems in the overall yield of product, the length of the route produces tremendous production scheduling difficulties to ensure an adequate supply in times of fluctuating market demand. The compounds of this series are nevertheless very valuable as a basis for mixtures intended for twisted nematic and supertwisted nematic displays because of their combination of wide nematic phase range and low viscosity. They also serve to illustrate an important theme in LC design — the systematic use of lateral fluorination to control smectic phase stability which is demonstrated by the compounds shown in Table 4.[19]

Many of the problems in the synthesis of novel LC species relate to the efficient assem-

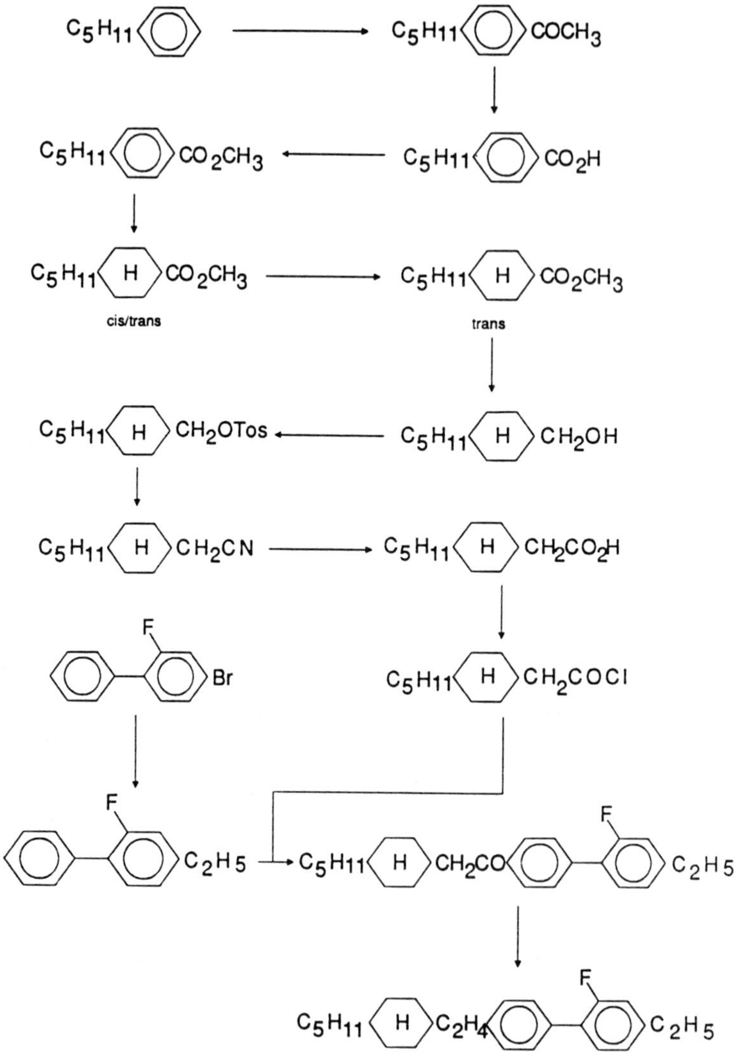

Figure 3: Synthetic Route to 'I' Series Liquid Crystals

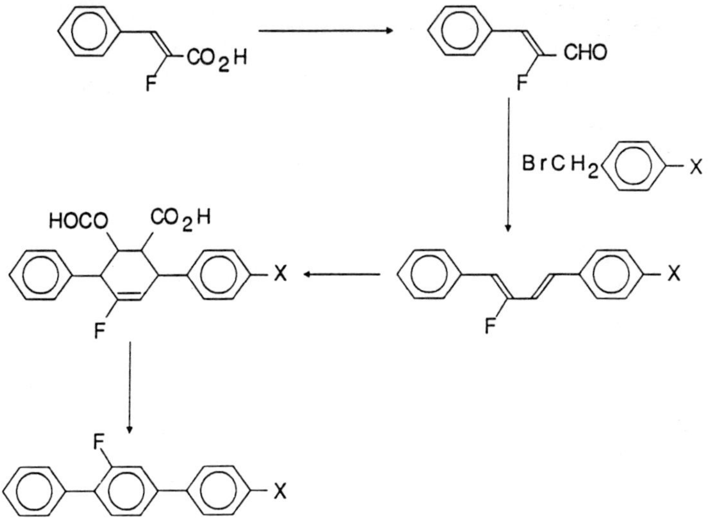

Figure 4: Early Route to Laterally Fluorinated Terphenyls

bly of 1,4-linked ring structures bearing lateral substituents in preferred positions. These difficulties have been greatly alleviated recently by the widespread application of aryl cross-coupling reactions, and this is illustrated in figures 4 and 5 which show two syntheses of the same laterally fluorinated terphenyl derivative. These syntheses each originate from the Hull University liquid crystal group, and date respectively from the late 1970's and from the latter part of the 1980's. The earlier route, employing successive Wittig coupling, Diels-Alder addition, decarboxylation and aromatisation gives a poor yield from an expensive starting material, while the latter is convenient to use, high yielding, and tolerant of a wide variety of substituent groups in position X and elsewhere in the molecule.[21]

The use of cross-coupling reactions has been particularly valuable in the synthesis of materials for ferroelectric displays. These devices offer an electro-optic effect which does not respond to an RMS voltage and which can be multiplexed to an indefinate level.[17,18] The response time, of the order of a few tens of microseconds in a multiplexed device, is also some three orders of magnitude shorter than in nematic based devices. These displays use a smectic C ($S_C$) phase in which the LC molecules are arranged in layers, and tilted with respect to the layer planes. Such phases are relatively uncommon, and when the electro-optic effect was first described in 1980 room temperature $S_C$ phases were almost unknown. The use of lateral fluorination once again provides a solution both by lowering the melting point of LC molecules and by modifying the nature of smectic phases to induce a tilt.[22,23] An extensive series of laterally fluorinated esters has been prepared in which the phase transition temperatures, birefringence and dielectric anisotropy can be varied

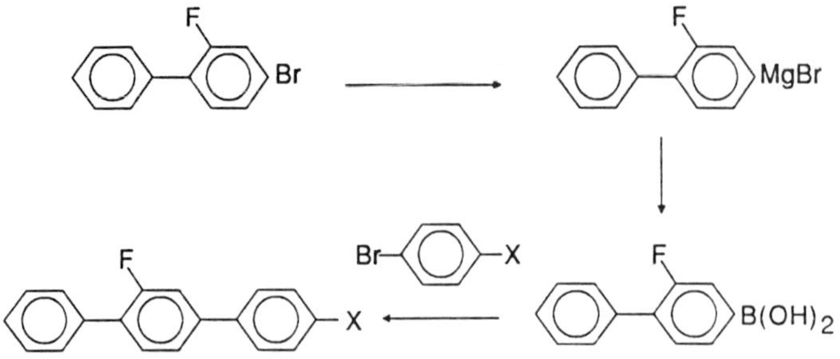

Figure 5: Improved Route to Fluorinated Terphenyl Synthesis

via the substitution pattern (Table 5). The fluorinated terphenyls also shown in Table 5 are particularly interesting, however, because the absence of a dipolar ester group helps to maintain a low viscosity and fast response time for terphenyl based mixtures.[21,24]

No ferroelectric properties can be observed in $S_C$ materials in the absence of a chiral dipolar group or additive. It is desirable for a strong lateral dipole to be coupled tightly to the chiral centre to obtain maximum spontaneous polarisation: there are many other demands on such a dopant such as solubility in the $S_C$ host, and the need to avoid inducing a very short pitch helical structure in the final composition. A very useful series of compounds are the cyanohydrin esters in Table 6 which impart ferroelectric polarisation to $S_C$ hosts at low concentrations.[22]

The future status of ferroelectric displays remains uncertain: complex devices with impressive performance have been demonstrated, but the displays have remaining problems of physical robustness and operating temperature range. If these can be overcome, ferroelectric displays seem the most promising candidate for very high resolution flat panel displays in applications such as workstations.

The final solution to the problem of addressing a complex format display can be found in the use of an active backplane, on which a separate switching element is fabricated to control each pixel. The limitations of RMS addressing no longer apply and the provision of greyscale capability can be left to the electronics. Nevertheless, the use of active drivers does not leave the LC material designer without a challenge. The operation of an active matrix backplane applies a pulse of voltage to each pixel to switch it to the desired state; the driving element then switches off for the remainder of the frame time of the device and the electric charge has to be stored on the pixel until it is refreshed on the next addressing cycle. The limited power available also means that low voltage operation is required.

Low voltage operation of a twisted nematic device may be related through the dielectric

| Structure | Phase transitions |
|---|---|
| C7H15CO2–[Ph]–[Ph(F,F)]–OC8H17 | K 54.9 S(C) 63.0 N 66.5 I |
| C7H15–[Cy]–CO2–[Ph(F,F)]–[Ph]–OC8H17 | K 54 S(C) 96 N 163.1 I |
| C7H15–[Cy]–CO2–[Ph]–[Ph(F,F)]–OC8H17 | K 53 S(C) 132.8 S(A) 144.4 N 162 I |
| C7H15–[Cy]–CO2–[Ph]–[Ph(F,F)]–C7H15 | K 35 S(B) 59 S(C) 79.3 S(A) 128 N 142.8 I. |
| C5H11–[Ph]–[Ph(F,F)]–[Ph]–OC8H17 | K 48.5 S(C) 95.1 N 141.5 I |
| C5H11–[Ph]–[Ph(F,F)]–[Ph]–C7H15 | K 36.5 S(C) (24) N 110 I |
| C5H11–[Ph]–[Ph]–[Ph(F,F)]–OC8H17 | K 85.8 S(C) 144.1 S(A) 147.9 N 158.3 I |
| C5H11–[Ph]–[Ph]–[Ph(F,F)]–C7H15 | K 55 S(C) 105 S(A) 130 N 135 I |

Table 5: Liquid Crystals for Smectic C Applications

| Structure | MPt | Spontaneous polarisation |
|---|---|---|
| $C_8H_{17}O$—⟨○⟩—⟨○⟩—$CO_2\underset{CN}{C}HCH(CH_3)_2$ | 67 | 170 |
| $C_8H_{17}O$—⟨○⟩—⟨○⟩—$CO_2\underset{CN}{C}HCH_3$ | 98 | 170 |
| $C_8H_{17}O$—⟨○(F)⟩—⟨○⟩—$CO_2\underset{CN}{C}HCH(CH_3)_2$ | 50 | 250 |
| $C_8H_{17}O$—⟨○(F)⟩—⟨○⟩—$CO_2\underset{CN}{C}HCH_3$ | 67 | 190 |

Table 6: Chiral Dopants for Smectic C Liquid Crystals

anisotropy to the dipolar nature of the LC molecules:

$$V_C^2 = \pi^2 K_{11}/\varepsilon_0 \Delta\varepsilon$$
$$\Delta\varepsilon = (NhF/\varepsilon_0)[\Delta\alpha - (F\mu^2/2k_BT)(1 - 3\cos^2\omega)]S$$

A full discussion of all the terms in these equations may be found in standard texts on liquid crystals; it is sufficient for the present to note that the dominant term in the second equation is that containing the molecular dipole moment $\mu$ and the angle made between this dipole and the molecular long axis $\omega$. The expected dielectric anisotropy rises as the square of the dipole moment. It is standard practice in the design of LC materials for twisted nematic displays to employ terminal substituent groups of a strongly dipolar nature to induce the required high positive dielectric anisotropy. The relative effectiveness of different candidate substituent groups may be predicted from the dipole moments of model structures such as the substituted benzenes of Table 7.

In practice, it is the cyano group which has found overwhelmingly the greatest application in liquid crystal materials for display use since the discovery of the cyanobiphenyls in the early 1970's. This is a consequence of the ability of the cyano group to provide highly stable, colourless nematic liquid crystals.

It is an experimentally observed fact, that the above equations fail to account for the variation of dielectric anisotropy with structure in real liquid crystals. Deviations from the expected behaviour are attributed to the tendency for terminal polar substituted liquid crystal molecules to form transient dimeric species. The preferred configuration for these

| X | $\mu$ |
|---|---|
| $CH_3$ | 0.36 |
| $F$ | 1.58 |
| $CF_3$ | 2.86 |
| $CN$ | 3.2 |
| $NO_2$ | 4.27 |

Table 7: Dipole Moments of Some Substituted Benzenes

dimers is with the dipolar groups arranged antiparallel, leading to an effective cancellation of the dipole and consequent reduction in the observed dielectric anisotropy of the compound.[25] A very few liquid crystals are known in which *parallel* dimers are preferred and which can show very high values of anisotropy. The esters shown below are a notable example

R—⟨○⟩—CO$_2$—⟨○⟩—CN
  (with F on second ring)

of compounds which show this behaviour. The underlying factors which determine the degree and nature of dimer formation in liquid crystals are porly understood: elucidation of the forces involved would provide a valuable input into the molecular engineering of new nematogens as well as providing a challenge for physical chemists.

In active matrix displays, in which low voltage operation is problematic, it might be assumed that mixtures containing the most highly dipolar substances available would be widely used. This turns out to be incorrect because of the other constraints on the properties of the mixture used, which turn out to be most severe in the areas of resistivity and stability. This arises from the addressing technique used in active backplane devices described above. In order for a display panel to show a high quality image, each pixel must be able to store the electric charge applied to it for the duration of the frame address time of the device (typically 20ms) with a tolerance corresponding to a single grey level on the device. This must be achieved not only under laboratory conditions, but throughout the anticipated lifetime of the device and under conditions of high operating temperature after exposure to ambient UV radiation, high humidity etc.

Recent work has shown that suitable liquid crystals based on highly stable core structures substituted with fluorinated groups are for the first time able to offer RC time constants of the order of 10's of seconds even after exposure to conditions of accelerated lifetime testing. These materials therefore offer the promise of fulfilling the needs for liquid crystals for very complex high quality active matrix displays towards the latter part of the 1990's.[26]

At the end of this decade, we reach also the end of the century and the millenium. It will be a rash man indeed who dares predict the course of display technology or its

| Structure | Dielectric Anisotropy | Viscosity | N - I |
|---|---|---|---|
| C$_5$H$_{11}$–[Cy]–[Ph]–F | 3 | 3 | |
| C$_5$H$_{11}$–[Cy]–[Ph]–CF$_3$ | 11 | 3 | |
| C$_5$H$_{11}$–[Cy]–[Ph]–OCF$_3$ | 7 | 4 | |
| C$_5$H$_{11}$–[Cy]–[Cy]–[Ph]–CF$_3$ | 9 | 28 | 123 |
| C$_5$H$_{11}$–[Cy]–[Cy]–[Ph](F)–F | 10 | 30 | 105 |
| C$_5$H$_{11}$–[Ph](F)–[Ph]–C$_2$H$_4$–[Ph]–OCF$_3$ | 11 | 25 | 82 |

Table 8: Fluorinated Nematic LC's for Active Matrix Use

attendant chemistry over such a timescale. Possibly, however, their prognostication might be aided by the innovative device shown in Figure 6. Whatever the other technical difficulties encountered in its construction, the author believes that the best candidate for the display in such equipment will remain a liquid crystal device for a number of years to come.

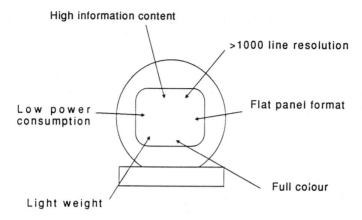

Figure 6: New Generation Crystal Ball

REFERENCES

1. T. Nagayasu et. al., Proc Int Disp Res Conf, 1988, 56
2. S. Komura et. al., Proc Japan Disp Conf, 1989, 528
3. H. Yamazoe et. al., Proc Japan Disp Conf, 1989, 536
4. A. Yasuda, H. Mori and J. Seto, J Appl Electrochem, 1987, 17, 567
5. A. Yasuda and J. Seto, J Electroanal Chem, 1988, 247, 193
6. C. Waters, V. Brimmell and E. P. Raynes, SIM Symp Digest, 1984, 261
7. N. Clark and S. Lagerwall in 'Liquid Crystals of One and Two Dimensional Order', ed. W. Helfrich and G. Heppke, Springer Verlag, Berlin, 1980
8. J. F. Clerc et. al., Proc Japan Disp Conf, 1989, 188
9. M. Schadt and W. Helfrich, Appl Phys Lett, 1971, 18, 127
10. J. Fergason, Proc SID Conf, Orlando, 1985
11. G. Anderson et. al., Appl Phys Lett, 1987, 51, 640
12. S. LeBerre et al, Displays, 1981, 349
13. G. Heilmeier, L. Zanoni and L. Barton, Appl Phys Lett, 1968, 13, 46
14. I. Sage in 'Thermotropic Liquid Crystals', ed G. Gray, Wiley and Sons, Chichester, 1987
15. K. Toyne in 'Thermotropic Liquid Crystals', ed G. Gray, Wiley and Sons, Chichester, 1987
16. P. Alt and P. Pleshko, IEEE Trans Electron Devices, 1974, 21, 146
17. P. Ross, Proc Int Disp Res Conf, 1988, 185
18. H. Inoue et. al., Proc Int Disp Res Conf, 1988, addendum
19. P. Balkwill, D. Bishop, A. Pearson and I. Sage, Mol Cryst Liq Cryst, 1985, 123, 1
20. B. Sturgeon, Phil Trans R Soc Lond, 1983, A309, 231
21. G. Gray, M. Hird, D. Lacey and K. Toyne, J Chem Soc Perkin Trans II, 1989, 2041
22. D. Bishop, J. Jenner, I. Sage, Proc 11th Int LC Conf, Berkeley, 1986
23. M. Chambers et. al., Liq Cryst, 1989, 5, 153
24. D. Coates, I. Sage, W. Crossland and A. Davey, Proc Japan Disp Conf, 1989, 176
25. K. Toriyama and D.Dunmur, Mol Cryst Liq Cryst, 1986, 139, 123
26. H. Plach, G. Weber and B. Rieger, Proc Int Disp Conf, 1990, 91

# The Physics of Displays for the 1990s


E. P. Raynes

ROYAL SIGNALS AND RADAR ESTABLISHMENT, MALVERN,
WORCESTERSHIRE WR14 3PS, UK


1  INTRODUCTION

During the last ten years our industrialised society has become characterised by the generation and flow of information, and it has become increasingly necessary to be able to display this information directly to the human operator. Traditionally the cathode ray tube (CRT) has been the standard means of displaying information in both domestic TV and business computing applications. Various requirements have demanded alternative display technologies to the CRT, and during the 1980's, liquid crystal displays (LCD's) have become prominent by virtue of their low voltage operation, low power consumption, design flexibility, and flatness. LCD's of low to moderate complexity have become a well established consumer item in the last few years, with current annual production of around $10^9$ units. However, LCD's with high complexity, for example computer or TV screens with over $10^6$ picture elements, are now becoming available using a variety of LCD technologies. We will now discuss the basic constructional principles and physics of current LCD's and see how they can be extended to the high complexity displays required for the 1990's.

2  PHYSICS OF NEMATIC LIQUID CRYSTALS

In this section we will describe the phase structures found in liquid crystals, their important physical properties, their alignment both on surfaces and in an electric field, and the principles behind electro-optic devices.

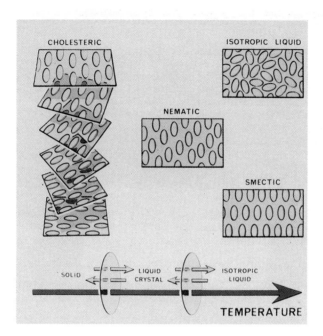

**Figure 1** Phase structure of thermotropic liquid crystals.

## Phase Structures Found in Liquid Crystals

The liquid crystal (LC) materials used in display devices are all thermotropic - they show an LC phase over a range of temperature. There are three main divisions of LC phases shown schematically in Figure 1.

*Nematic*. The nematic phase (N) is the most widely used LC phase and possesses orientational order of the molecular axes.

*Cholesteric*. The cholesteric phase (N*) is closely related to the nematic phase, with a helical ordering imposed so that rotation of the ordering direction occurs over a pitch length P which can be as short as 0.1 µm.

*Smectic.* In contrast to the N and N* phases, smectic (S) LC's also show various degrees of positional ordering of the centres of gravity of the molecules into layers.[1]

It should be noted that neither the orientational nor the positional ordering into layers is perfect as suggested by Figure 1, which is simplistic and illustrates the principles of the ordering. The reality is best described as a tendency for the molecules to align in one direction or into layers.

## Physical Properties of Nematic Liquid Crystals

Nematic liquid crystals are anisotropic uniaxial liquids with two principal values of the physical properties. Foremost amongst these are the refractive indices, the electric permittivities and the elastic constants.

*Refractive Indices*. The ordinary ($n_o$) and extrordinary ($n_e$) refractive indices are different, making nematic LC's into birefringent liquids. The birefringence ($\Delta n = n_e - n_o$) is an important quantity determining the optical properties of thin layers of nematic LC's. Current LC materials show a range of $\Delta n$ from +0.03 to +0.40.

*Electric Permittivities*. There are two electric permittivity components, $\varepsilon_\parallel$ measured along the average molecular axis, and $\varepsilon_\perp$ perpendicular to this axis. The magnitude and sign of the anisotropy of the electric permittivity ($\Delta \varepsilon = \varepsilon_\parallel - \varepsilon_\perp$) determines the response of a nematic LC to an applied electric field. Permittivity anisotropies for known LC materials lie within the range $-10 < \Delta \varepsilon < +50$, and are dominated by the presence of dipole moments within the molecules.

*Orientational Elasticity*. The average direction of the molecular axes is descibed by a unit vector known as the director ($\underline{n}$), which prefers to be everywhere parallel within the LC layer. Distortion of this alignment by electric fields produces an energy of distortion first written down by Frank [2] which involves three elastic constants; $K_{11}$ (splay), $K_{22}$ (twist) and $K_{33}$ (bend). All three elastic constants are small ($\approx 10$ pN), but their magnitudes and ratios are crucial in determining the electro-optic properties of nematic LCD's.

## Alignment of Nematic Liquid Crystals

The orientation of liquid crystals on surfaces is fundamental to the construction of display devices. Two principal alignments exist.[3] The normal, or "homeotropic", alignment can be achieved with a variety of surfactants, and the planar, or "homogeneous", alignment can be produced either by oblique evaporation of a dielectric layer, or by buffing a previously deposited polymer coating. The latter technique is simple to implement and is employed almost universally, using thin layers of polyimide.

The alignment of a nematic LC by an electric field is a direct result of the interaction with the anisotropic electric permittivity and the minimisation of the electrostatic energy. Nematics with positive dielectric anisotropy ($\Delta\varepsilon > 0$) want to align parallel to the applied electric field, and those with negative anisotropy ($\Delta\varepsilon < 0$) perpendicular to the field.

## 3 LIQUID CRYSTAL ELECTRO-OPTIC EFFECTS

### Principles of LC Electro-optic Effects

Electro-optic effects in LC's are produced by confining and aligning a thin layer of LC between two conducting glass plates which have been treated to produce surface alignmnent. The alignment of the LC layer can then be changed by the application of a voltage between the two plates; removal of the voltage restores the original alignment determined by the two glass plates. The optical properties of thin LC layers depend on the molecular orientation within the layers. Application of a voltage across the cell therefore results in a reversible electro-optic effect.

### Basic Liquid Crystal Display Construction

The basic LCD construction, shown in Figure 2, is described in more detail by Clark et al.[4] An LCD consists of a thin layer, of between 2 - 10μm, of liquid crystal

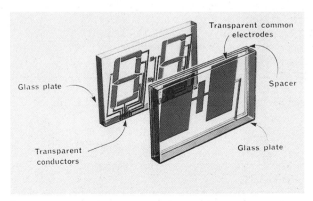

Figure 2  Construction of a simple LCD

material, confined between two glass plates coated with a transparent conductor which has been selectively etched using standard techniques to produce the desired electrode pattern. After rigorous cleaning the inner glass surfaces are treated to produce the required alignment using a combination of chemical and mechanical techniques.[3] A variety of heat or UV sensitive epoxy-based adhesives are then used to seal the glass plates together with a spacing determined by glass fibre or plastic ball spacers distributed across the cell area. The liquid crystal is introduced, using an evacuated chamber, through a gap in the seal which is subsequently closed using epoxy resin.

### The Twisted Nematic Electro-optic Effect

The twisted nematic (TN) device shown in Figure 3 is the most widely used LCD. It was first reported by Schadt and Helfrich,[5] and uses an optical effect reported many years earlier by Mauguin.[6] The device uses two orthogonally aligned surfaces which produce a 90° twist in the molecular axes, guiding, or rotating, the plane of polarisation of

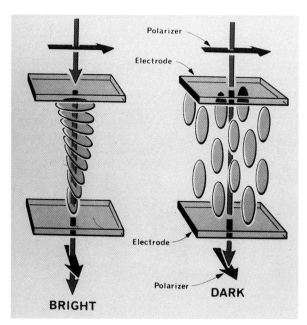

Figure 3  Operation of a Twisted Nematic LCD

light polarised parallel or orthogonal to the surface alignment direction. The application of around 2 to 3 volts produces realignment of the molecules, such that the optical guiding property is lost. Placing the device between crossed or parallel polarisers therefore produces a high contrast electro-optic switch.

The detailed optics of twisted nematic layers can be calculated using the Jones Matrix formalism,[7] which neglects any reflected light, a reasonable approximation for twisted nematic layers. The Jones Matrix method shows that the transmission of light of wavelength $\lambda$ through a 90° twisted nematic of thickness d, using parallel polarisers, is given by:[8]

$$T = \frac{\sin^2 \pi/2 \sqrt{\{1 + (2\Delta n.d/\lambda)^2\}}}{1 + (2\Delta n.d/\lambda)^2} \qquad (1)$$

The transmission oscillates, going through minima for specific values of $(\Delta n.d/\lambda)$, and TN devices therefore require careful choice and control of both $\Delta n$ and d.

## 4  NEMATIC LIQUID CRYSTAL DISPLAYS FOR THE 1990's

The displays of the 1990's will be far more complex than those of the 1980's. They will possess over 1000 X 1000 picture elements, as well as having full colour and grey scale capabilities. Although today's TN LCD's are ideal for many applications, they are unable to display a large amount of information without either the inclusion of electronic switching elements, or a major modification to the strucure of the display. The limitations of the TN display result from the gentle change of transmission with voltage found in them. Even when optimised driving schemes are used,[9] and the device construction and LC material are both optimised,[10] performance is still inadequate. There are two quite separate developments of nematic LCD's which have the prospect for taking nematic LCD's into complex displays for the 1990's.

### Active Matrix Displays

The integration of an active nonlinear electronic device into the display adjacent to the liquid crystal layer is a powerful technique that greatly improves the performance of complex TN LCD's. The complexity now resides in the electronic device, which must permit the capacitance of the liquid crystal to charge up rapidly, and then prevent the charge leaking away whilst the rest of the matrix is being addressed.

Several different types of nonlinear device are currently being exploited. Foremost amongst these are metal-insulator-metal films (MIM), field effect transistors (FET) fabricated in single crystal silicon, and thin film transistors (TFT) fabricated using a variety of materials. Displays using FET's fabricated in single crystal silicon are restricted in size, and most attention recently has been devoted to TFT's using CdSe,[11] amorphous silicon,[12] or recrystallised poly-silicon.[13] Small, full colour TFT LCD's with grey scale are now available commercially for the portable consumer TV market, and larger TFT displays are available, at a price, for the professional market. There are significant production difficulties of these devices, particularly the large area ones, but these problems all lie in the successful fabrication of large arrays of these solid state devices, not in the LCD itself.

Supertwist Displays

A dramatic improvement in the performance of large, complex, nematic LCD's occurred in 1982, when it was observed that the voltage dependence of the transmission of nematic LC layers with twist angles in the range 180° to 270° could become infinitely steep.[14] This is illustrated by the transmission/voltage curves for the common nematic mixture ZLI 1132 in Figure 4. The larger twist angles are produced by a combination of surface alignment and making the nematic mixture into a long pitch cholesteric by the addition of a small amount of a chiral twisting agent. The increasing
twist angle steepens the transmission/voltage curve, until it becomes bistable for 270° twist; for a specific twist angle between 225° and 270° the curve becomes infinitely steep and well suited to multiplexing. The larger twist angles present have resulted in the name supertwisted nematic (STN) for these LCD's.

Much of the physics of STN displays is now understood, and methods exist for designing both the device and the material to give optimum performance. One of the problems with the early STN displays was the colouration present which prevented the fabrication of a black/white device. The optical properties of STN layers can be calculated using the Jones Matrix method which was so suitable for the

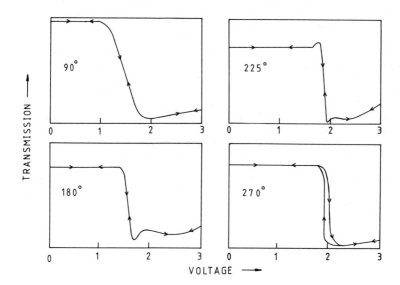

**Figure 4** Voltage dependence of the transmission of highly twisted LCD's constructed using ZLI 1132

TN device. Application of the method[15] gives analytical results for the wavelength dependent transmission. For example, a layer with 270° twist and crossed analyser and polariser at 45° to the alignment direction, has a transmission given by:

$$T = \cos^2 3\pi/2 \sqrt{\{1 + (2\Delta n.d/3\lambda)^2\}} \qquad (2)$$

Equation 2 gives a transmission spectrum with a peak at $\lambda = 2\Delta n.d/\sqrt{7}$ and a width which is narrower than the visible spectrum. The display is therefore coloured, although the colour can be changed by varying $\Delta n.d$. Several methods now exist for removing this colour and producing a black/white device. The most elegant method uses a second STN layer[16] to compensate the colouration produced by the first layer. The second layer must be identical to the first except that the twist angle is in the opposite sense, and if the second layer is offset by 90° to the first, then exact optical compensation occurs. This can be readily proved using Jones Matrices.[17] The use of a second layer to compensate the colouration is however an expensive solution, and an alternative technique based on birefringent plastic layers[18] is now becoming widely used. Once the STN display has been made into a black/white device it is also possible to add colour filters to give full colour displays.

STN displays are also inherently slower than TN LCD's, with response times of around 200ms. They are therefore too slow for video applications, or for computer terminals with rapidly moving cursors, although developments are taking place using thinner LC layers to reduce the response times to 50ms. Finally, grey scale is also a problem; the STN display is close to bistability, making grey scale difficult to achieve, although limited grey scale is now being claimed by some manufacturers.

## 5 FERROELECTRIC DISPLAYS FOR THE 1990's

The possibility of the existence of ferroelectricity in chiral, tilted smectic liquid crystal phases, and of devices based on it, was first predicted in 1975 by Meyer.[19] Suitable systems based on smectic C liquid crystals were synthesised within a year and shown to be ferroelectric. By 1980 the first practical device using the ferroelectric phase was demonstrated,[20] and since then research towards complex ferroelectric displays has advanced rapidly, opening up many fascinating new areas of physics and chemistry.

### Origin of Ferroelectricity in Chiral Smectic C Phases

The molecular origin of ferroelectricity in chiral, tilted smectic phases is illustrated in Figure 5. This shows an archetypal ferroelectric $S_C^*$ LC molecule with a dipole moment residing on a chiral centre on one end of the molecular core. The middle layer in Figure 5 shows all possible orientations of the $S_C^*$ molecules; clearly two of the orientations are a better fit within the smectic layer, and these are shown in the lower diagram. This suggests a hindrance of the molecular rotation and a net dipole moment which is positive in the example shown. Tilt to the left, or the use of the opposite chirality would result in a negative spontaneous polarisation.

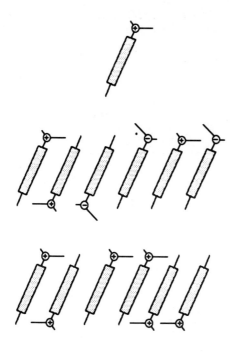

Figure 5  The hindrance of molecular rotation and the resulting spontaneous polarisation in $S_C{}^*$ layers

## Ferroelectric Liquid Crystal Materials

There has been considerable progress in ferroelectric materials over the past few years. It has become standard practice to form an $S_C$ host mixture, and add to it a chiral dopant to impart ferroelectricity.

The di-fluoroterphenyls[21] (di-FTP) represent a good example of a recent materials development. Simple mixtures based on the two types of di-FTP's shown in Figure 6 can possess a wide $S_C$ range from well below 0°C up to 100°C. The addition of the chiral dopant[22] shown in Figure 6 results in a stable, wide temperature range, ferroelectric mixture which exhibits 1µs switching in room temperature devices.

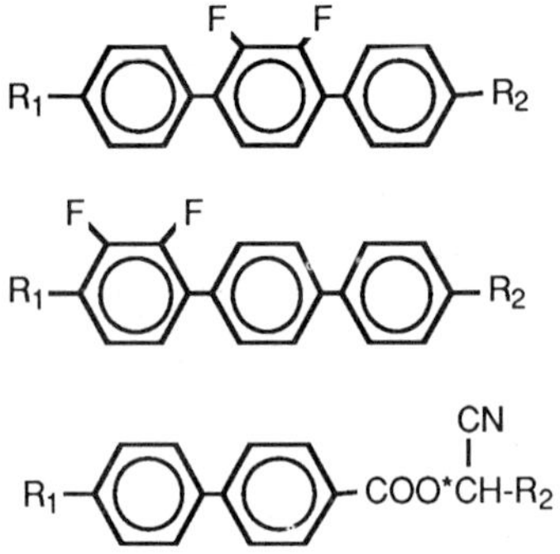

Figure 6  Some recent ferroelectric materials

Ferroelectric Liquid Crystal Devices

Although first recognised by Meyer,[19] the potential of ferroelectric LCD's as fast bistable devices was not demonstrated until 1980.[20] Alignment of both the molecular axes and the smectic planes is vital for good device performance and can be achieved by a combination of standard surface alignment techniques and cooling through the phase sequence

$$I - N^* - S_A - S_C^*$$

together with an elongated $N^*$ pitch.[23]

$$P > 4d$$

The alignment process, illustrated in Figure 7, results in two smectic C states, and the application of a DC electric field switches the layer between the two states. Figure 8 illustrates the mode of operation of a ferroelectric display using suitably oriented crossed polarisers. Switching can be achieved within 1μs at room temperature with the best materials and together with the bistability makes ferroelectric displays very attractive for fast electro-optical devices and for complex displays.

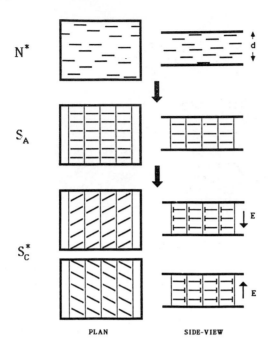

Figure 7  Alignment of $S_C^*$ by a combination of phase sequence and elongated $N^*$ pitch

## New Physics of Ferroelectric LCD's

There are several new areas of physics within the field of ferroelectric liquid crystals which are currently becoming understood. We will now discuss three of these.

Alignment of the $S_C^*$ Layers. The simple picture of alignment given above is adequate for the $S_A$ phase, but is inconsistent with all the known facts of the $S_C^*$ phase. The parallel layers in Figure 7, known as the "bookshelf geometry", are incompatible with the surface alignment (molecular axes approximately parallel the surface and in a specific direction) and the normal $S_C^*$ cone angle of ≈22°. X-ray studies[24] have recently shown that the smectic layers tilt on cooling from the $S_A$ to the $S_C^*$ phase, adopting the

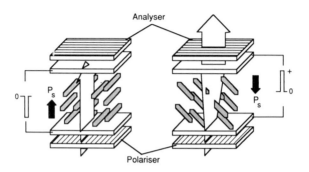

Figure 8  The two bistable switched states of a thin layer of $S_C^*$ liquid crystal

"chevron" structure illustrated in Figure 9. The layers tilt at an angle which is approximately 2° less than the $S_C^*$ cone angle. Chevrons of opposite directions are seperated by very characteristic defects known as "Zig-zags".

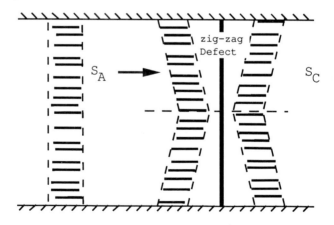

Figure 9  The chevron structure adopted by $S_C^*$ layers and the zig-zag defect

Alignment of the Molecular Axes. Any profile of the molecular axes across the cell must be compatible with the chevron structure of the layers. In the centre of the cell, the continuity of the molecular axis between the two halves of the chevron results in a small twist angle of ≈8° away from the surface alignment direction, which is ≈0°. The simplest profile is then the "Twisted Director Profile" (TDP),[25] in which the molecular axis twists out by ≈8° towards the centre of the layer, and then back again towards the other surface. Optically the layer is like two 8° TN layers in series, and once again the Jones Matrix method is a useful method of analysing the optical

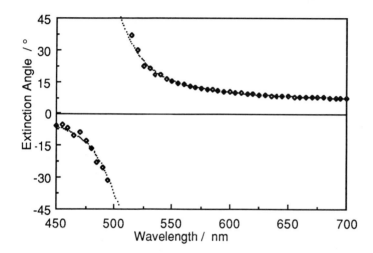

Figure 10 The wavelength dependent extinction angle of an $S_C$ layer; the dots show the theoretical curve fitted to the experimental data points

properties.[25] Figure 10 shows the comparison between experimental data and the results of the Jones Matrix calculation on the wavelength dependent extinction angle of an $S_C$ layer.[25] The excellent agreement of the experimental data and theory provides strong support for the validity of the TDP.

Biaxiality. Uniaxial phases such as the N and $S_A$ phases can show two principal values of physical properties. The ferroelectric $S_C^*$ phase is biaxial, and can therefore show three principal values. Early measurements[26] showed that any biaxiality of the refractive indices was small. However, it was shown in 1989 [27] that the electro-optic properties of certain ferroelectric LC cells could only be explained by assuming significant biaxiality of the electric permittivities. The actual biaxial permittivities were measured,[28] and Figure 11 shows values for a diFTP mixture.

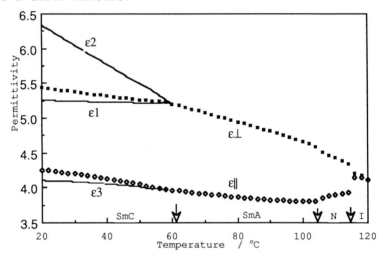

Figure 11  The biaxial permittivities of a diFTP mixture

Electric Field Effects. Nematic liquid crystals respond to an electric field via the anisotropic permittivity, resulting in an energy term $\Delta\varepsilon.E^2$ which can reorientate the molecular axis when an electric field is applied. A ferroelectric liquid crystal also has an energy term related to the spontaneous polarisation $P_S.E$, and therefore the total interaction driving the field induced molecular reorientation is given by:[29]

$$\{ P_S.E \ + \ \Delta\varepsilon.E^2 \} \tag{3}$$

The $P_S.E$ term switches the cell into the other state for the correct polarity of E. If $\Delta\varepsilon < 0$, which is the case for most $S_C^*$ materials, then an applied field of either sign stabilises the existing state. For a low field, the cell switches to the other state; conversely, for a large field, the $\Delta\varepsilon.E^2$ term dominates and the cell does not switch. There also exists an intermediate field which produces the minimum response time,[30] and the magnitude of this field is proportional to the ratio $P_S/\Delta\varepsilon$.[31] The presence of dielectric biaxiality, $\partial\varepsilon$, will replace $\Delta\varepsilon$ by some combination of $\Delta\varepsilon$ and $\partial\varepsilon$, but the arguments leading to the presence of a minimum in the response time remain valid.

## 6 CONCLUSIONS

LCD's have been a major force in the display industry of the 1980's; it seems certain that their dominance will not only continue into the 1990's, it will probably grow further. The trend into the 1990's will be towards more complex displays, and there are at least three options using liquid crystals. Two of these are based on extensions of the existing nematic LCD's, whereas the third is based on the very new, and relatively poorly understood, devices based on ferroelectric liquid crystals. It is too early to judge the winner out of these three options; indeed all may be successful for different applications.

### ACKNOWLEDGEMENTS

Copyright © Controller HMSO London, (1990)

### REFERENCES

1. G.W. Gray and J.W. Goodby, 'Smectic Liquid Crystals', Leonard Hill, Glasgow and London, 1984.
2. F.C. Frank, Discuss. Faraday Soc., 1958, 25, 19.
3. E.P. Raynes, 'Electro-optic and Photorefractive Materials' ed. P. Gunter, Springer-Verlag, Berlin Heidelberg, 1987, Part II, 80-109.
4. M.G Clark, K.J. Harrison and E.P. Raynes, Phys. Technol., 1980, 11, 232.
5. M. Schadt and W. Helfrich, Appl. Phys. Lett., 1971, 18, 127.
6. C. Mauguin, Bull. Soc. Fr. Miner., 1911, 34, 71.
7. E.P. Raynes and R.J.A. Tough, Mol. Cryst. Liq. Cryst. Lett., 1985, 2, 139.
8. C.H. Gooch and H.A. Tarry, J. Phys. D, 1975, 8, 1575.
9. P.M. Alt and P. Pleshko, IEEE Trans. Electron. Devices, 1975, ED-21, 146.
10. E.P. Raynes, IEEE Trans. Electron. Devices, 1979, ED-26, 1116.

11. F.C. Luo, W.A. Hestor and T.P. Brody, Society for Information Display Digest, 1978, 94.
12. A.J. Snell, K.D. Mackenzie, W.E. Spear, P.G. LeComber and A.J. Hughes, Appl. Phys., 1981, 24, 357.
13. T. Nishimura, Y. Akasaka, H. Nakata, A. Ishizu and T. Matsumaoto, Society for Information Display Digest, 1982, 36.
14. C.M. Waters, V. Brimmell and E.P. Raynes, Proc. 3rd. Int. Display Res. Conf., Kobe, Japan, 1983, 396.
15. E.P. Raynes, Mol. Cryst. Liq. Cryst. Lett., 1987, 4, 69.
16. K. Katoh, Y. Endo, M. Aktsuka, M. Ohgawara and K. Sawada, Jap. J. Appl. Phys., 1987, 26, L1784.
17. M.J. Towler, (private communication).
18. S. Matsumoto, H. Hatoh and A. Murayama, Proc. 12th Int.Liq. Cryst. Conf., Freiburg, Germany, 1988, 349.
19. R.B. Meyer, Mol. Cryst. Liq. Cryst., 1977, 40, 33.
20. N.A. Clark and S.T. Lagerwall, Appl. Phys. Lett., 1980, 36, 899.
21. G.W. Gray, M.H. Hird, D. Lacey and K.J. Toyne, J. Chem. Soc., Perkin Trans.2, 1989, 2041.
22. M.J. Bradshaw, V. Brimmell, J. Constant, J.R. Hughes, E.P. Raynes, A.K. Samra, L.K.M. Chan, G.W. Gray, D. Lacey, R.M. Scrowston, I.G. Shenouda, K.J. Toyne, J.A. Jenner and I.C. Sage, Proc. Soc. Inf. Display, 1988, 29/3.
23. M.J. Bradshaw, V. Brimmell and E.P. Raynes, Liquid Crystals, 1987, 2, 107.
24. T.P. Reiker, N.A. Clark, G.S. Smith, D.S. Parmar, E.B. Sirota and C.R. Safinya, Phys. Rev. Lett., 1987, 59, 2658.
25. M. Anderson, J.C. Jones, E.P. Raynes and M.J. Towler, submitted to J. Phys. D.
26. S. Garoff, PhD Thesis, Harvard University, 1977.
27. J.C.Jones, E.P. Raynes, M.J. Towler and J.R. Sambles, Proc. Brit. Liq. Cryst. Soc., Sheffield, UK, 1989, and Mol. Cryst. Liq. Cryst. Lett., in press
28. J.C. Jones and E.P. Raynes, Proc. Brit. Liq. Cryst. Soc., Bristol, UK, 1990.
29. Jiu-Zhi Xue, M.A. Handschy and N.A. Clark, Ferroelectrics, 1987, 73, 305.
30. H. Orihara, K. Nakamura, Y. Ishibashi, N. Yamamoto and M. Yamawaki, Jap. J. Appl. Phys., 1986, 25, 839.
31. F.C. Saunders, J.R. Hughes, H.A. Pedlingham and M.J. Towler, Liquid Crystals, 1989, 6, 341.

# Dye Diffusion Thermal Transfer Printing (D2T2)

R. A. Hann

ICI IMAGEDATA, BRANTHAM, MANNINGTREE CO11 1NL, UK

## 1 INTRODUCTION

All colour reproduction processes depend on the fact that the eye has three different types of colour sensor, which can be roughly characterised as being sensitive to red, green, and blue light. For example, a video camera has separate sensors for the red, green, and blue (RGB) parts of the spectrum, otherwise known as the three additive primary colours; these separate signals are normally encoded together for transmission, then decoded in the receiving television in order to give separate signals for driving the three electron guns that excite the red, green, and blue phosphors on the screen. In this way, there is a direct correspondence between, say, the red light level at a given point in the original scene and the red light emitted at the corresponding point in the image on the screen. It is interesting to reflect that if the human eye had had a significantly greater number of colour sensors, then the colour reproduction problem might very well have been too complex to tackle.

Colour print (or transparency) reproduction requires a further notional stage in the process, as it is necessary to reproduce the image using dyes or pigments that by their nature cannot emit light, but are only capable of absorbing it. In order to achieve reproduction, the three subtractive primary colours are used. These are Cyan (a greenish blue), which absorbs red light; Magenta (a dark pink), which absorbs green light; and Yellow, which absorbs blue light. The three subtractive primaries (CMY) can be regarded as the negative of the additive primaries, as illustrated in Figure 1. Thus, a white paper or clear transparency corresponds to the maximum light that can be passed by the system; if, for example we wish to reproduce a green colour then this must be done by absorbing red and blue light, by printing cyan and yellow dyes.

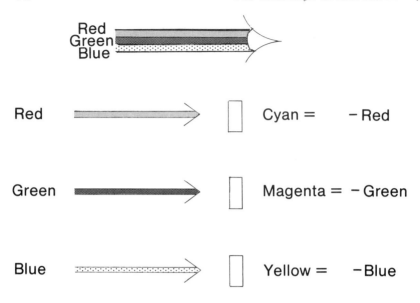

**Figure 1** The Relation Between Additive (RGB) and Subtractive (CMY) Primary colours

## 2 DYE DIFFUSION THERMAL TRANSFER PRINTING (D2T2)

Dye Diffusion Thermal Transfer printing (D2T2) is based on the transfer of dyes from a colour ribbon onto a receiver paper, using a thermal head as a source of heat. The thermal head is a hybrid circuit on an alumina substrate, the heat being generated by a line of resistive thin film tantalum elements[1]. Each of these elements is individually connected by thin film gold electrodes to logic circuitry elsewhere on the hybrid circuit, thus allowing the current through each heater to be turned on and off independently of its neighbours. The configuration of the heating elements is shown schematically in Figure 2. It will be noted that each element is split into two sections; this is to reduce the visibility of the line structure consequent on printing.

The quantity of dye transferred at each element depends on the length of the electrical pulse applied. It is therefore possible with this system to print a line at a time, with each element in the line having an appropriate depth of shade of the colour being printed, giving what is known as a continuous tone image.

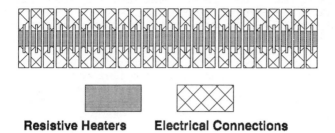

**Figure 2** Layout of the Heating Elements of a Thermal Head

The overall printing process is illustrated in Figure 3. The colour ribbon is arranged as a series of repeating blocks of colour (CMY), each of which is large enough to cover the image being printed. It is held in close contact with the thermal head by means of a rubber coated platen roller, and electrical signals are fed to it as the roller is slowly rotated. The complete image is thus printed in yellow, before being overprinted in magenta and then cyan to give the full colour picture.

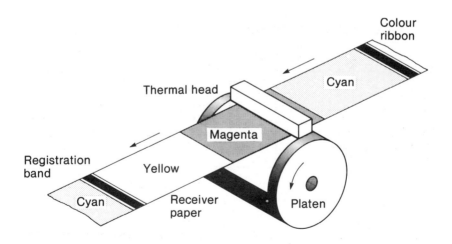

**Figure 3** D2T2 Printing (Schematic)

The printers are available from a number of manufacturers, and usually accept the colour ribbon in a cassette and the receiver paper as a flat pack. Most printers are capable of accepting an image either in digital form or as an analogue video image. In the latter case, the internal electronics of the printer digitises a single frame of the input signal and stores in memory the digitised brightness values for each point in the picture (typically 500 x 700 **pixels** for a video picture). There are three planes of memory, for the red, green and blue signals. In order to make the print, the data is read out of the memory, column by column, and the drive signals for the thermal head are obtained by using a look-up table to convert, for example, a low level of blue input to a long pulse time for the yellow printing cycle. This process is illustrated in Figure 4.

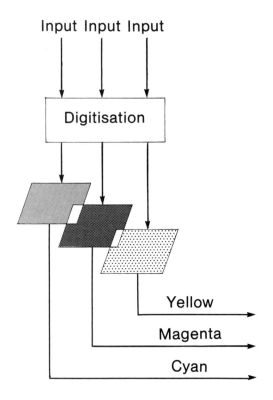

**Figure 4** Digitisation of Input RGB Signal and Conversion to CMY

The process has been referred to by some workers[2] as 'sublimation transfer printing'. This description is based on an analogy with sublimation printing of textiles. In that process, a pattern is first printed onto paper with aqueous inks made from disperse dyes. The printed paper is then placed in contact with the fabric to be dyed (usually polyester), and heat is applied. The dye sublimes, and the dye molecules diffuse through the vapour state, penetrate the gaps between the fibres, and provide all round dyeing. This is illustrated in Figure 5, where the shaded area represents the dye layer, and the dots represent dye molecules that are moving round the textile fibres (circles). In contrast, the D2T2 process involves direct contact between ribbon and receiver, which are held together under pressure; there is therefore no air gap for the dye molecules to sublime into, and the dye is transferred by a molecular diffusion process between the ribbon and the receiver.

## SUBLIMATION            DIFFUSION

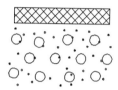    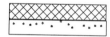

**Figure 5**  Contrast Between Sublimation and Diffusion Transfer

### 3 STRUCTURE OF THE MATERIALS

The structure of the materials is illustrated in Figure 6, which shows both the colour ribbon and the receiver paper. The colour ribbon is coated onto a base film of biaxially oriented polyester, typically of thickness 6 micrometres. The use of such a thin base film is determined by the requirement for the heat to pass through the film before it can interact with the dye layer. In order to protect the base film from the thermal head (which can reach temperatures in excess of 360 $^{o}$C (see later) and to allow slipping past the thermal head, a crosslinked backcoat is provided on the side nearest the head.

On the opposite side, a subcoat is used to provide a good key to the polyester surface and to provide a barrier to backwards migration of dye. The dye layer itself is deposited on top of this, and consists of a solid solution of dye in a binder polymer.

The receiver layer may be deposited onto a wide range of substrates, although in our present materials the preferred substrate is normally either clear 'Melinex' polyester or white 'Melinex' polyester. Both these materials are optically neutral, and the white 'Melinex' in particular provides an exceptionally accurate bright white background for the colour print.

The receiver layer has to receive the dye image and provide release after printing and is made by coating a soluble polyester and silicone mixture. The reverse side of the receiver has an anti slip coating in order to aid the printer feed mechanism.

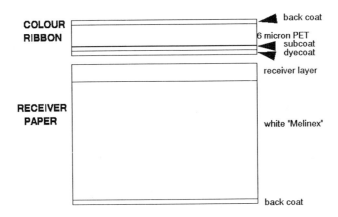

**Figure 6** Structure of Colour Ribbon and Receiver 'Paper'

It was recognised at an early stage of this programme that commercially available dyes would not be adequate for use as D2T2 colorants$^3$. A major dye synthesis and testing programme was therefore undertaken with ICI's Colours and Fine Chemicals Division in order to identify the structural features necessary for a good D2T2 dye and to provide the dyes used for the present materials. Some of the necessary performance requirements are listed below:

        Bright Colour (narrow absorption band).
        Strong Colour (high extinction coefficient).
        Easily transferred.
        Permanently fixed.
        Light fast.
        Non Toxic.

Some of these requirements are at first sight mutually exclusive; for example it seems unreasonable to expect a dye to be both easily transferable and permanently fixed. In practice a compromise has to be made on this pair of properties. Selecting the best dyes is thus an extremely complex problem that requires a deep level of understanding, as well as great skill and experience in dye chemistry. An initial screening process served to provide an outline understanding of the structural requirements for a good D2T2 dye, and led to some very useful empirical rules. Further progress then required a more detailed approach to understanding and modelling the transfer process.

## 4  MODELLING THE D2T2 PROCESS

During the printing process, the colour ribbon and receiver paper are held in intimate contact by means of pressure applied between the thermal head and the platen roller. The dye coat of the ribbon and the receiver layer of the paper are the faces actually in contact, and transfer of dye takes place in this defined area. As described above, the basic mechanism of transfer is by a diffusion process between two condensed phases.

The dye is originally in the form of a solid solution in the binder on the surface of the colour ribbon. As the temperature rises during printing, the dye equilibrates between the surface of the ribbon and the surface of the receiver, and partitions into the receiver polymer. This leads to a high concentration of dye just inside the receiver's surface, and this dye then migrates into the receiver polymer by simple Fickian diffusion. The dye migration process is indicated schematically in Figure 7. During the transfer the dye moves by a molecular diffusion mechanism, so that there is no bulk flow, and it remains dissolved either in the binder polymer or in the receiver polymer. There is therefore no need as well as no physical basis for a transfer mechanism based on sublimation.

Both the partition between the polymers and the diffusion of dye into the receiver are thermally activated processes[4]. The partition process has been measured for some of the dyes of interest and has been found to have the form given in Equation 1:

$$K = K_0 \exp(-E_K/RT) \qquad (1)$$

Where $K$ is the partition (distribution) coefficient of dye between binder and receiver polymers at absolute temperature $T$; $R$ is the universal gas constant, and $K_0$ and $E_K$ are the pre-exponential factor and activation energy determined from the measured data.

The diffusion coefficient of dye in the receiver can be expressed by a similar equation (2):

$$C = C_0 \exp(-E_C/RT) \qquad (2)$$

Where C is the diffusion coefficient, and the other factors have similar meanings to those in the previous equation.

It is clear that by combining this physical model with thermal data for the system, it should be possible to model the D2T2 process.

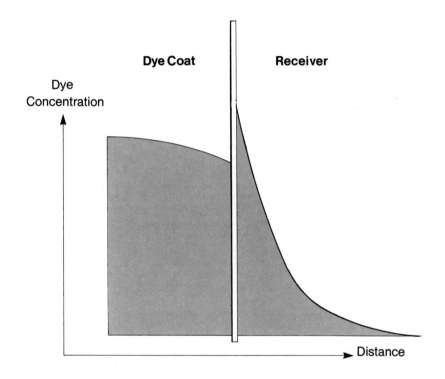

**Figure 7**  Dye concentration during thermal transfer (schematic).

*Dye Diffusion Thermal Transfer Printing (D2T2)* 155

**Thermal Model of The D2T2 Process**

The thermal head generates heat in defined areas of resistive elements (see Figure 2). This heat is lost by conduction back into the substrate, by lateral transmission in the head, and by dissipation into the sample; the heat flows have been modelled, thus allowing the surface temperature of the head to be determined, and more importantly providing information about the temperature distribution at the interface where dye transfer is taking place. The surface temperature of the head has also been measured independently by means of thermal microscopy. The results of this work will be described in more detail elsewhere[5]. Figure 8 shows the temperature distribution measured after 10 ms of head energisation.

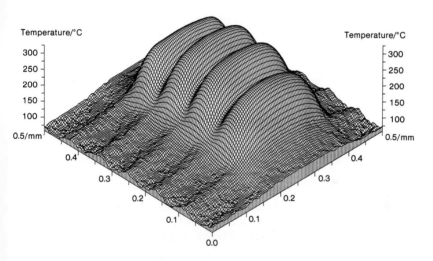

**Figure 8  Measured Temperature Distribution of the Surface of a Thermal Head (Two Elements Activated)**

One significant feature is the lower temperature to be observed in the outer halves of each element. This has observable consequences during printing, as it tends to lower the contrast of sharp edges; it also means that it is impossible to print an area of uniform colour by sequentially activating different parts of a line, because the edges of each block will always be too pale.

The calculated temperature distribution within the materials at various times after head energisation is shown in Figure 9. This implies materials temperatures of up to the region of 250 °C at the ribbon/receiver interface. One important consequence of these temperatures is that the key layers will be molten during the later stages of the dye transfer process. This emphasises the importance of having an efficient release system at the receiver's surface.

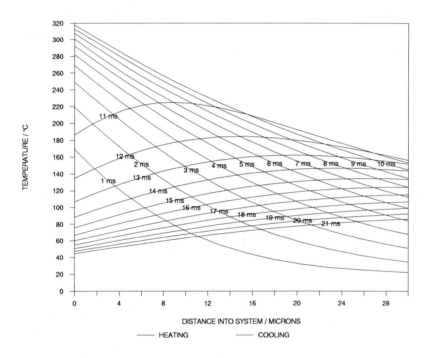

**Figure 9  Calculated Temperature Distribution within D2T2 Materials (Head energised for 10 ms, then Allowed to Cool)**

### Combining the Physical and Thermal Models

By combining the thermal model with the physical model and measured partition and diffusion coefficients, the dye transferred can be calculated at each point on the pixel area. Because of the higher temperatures near the centre of each heated element, more dye will be transferred in these regions, and there will thus be an optical density (OD) variation, with the highest densities at the pixel centre. In order to obtain an average value that can be compared with densitometer readings, it is necessary to average the reflected light over the whole area.

In the present work, this was done by calculating the expected transmission density ($d_T$) at each point by multiplying the predicted mass of transferred dye by the extinction coefficient determined from solution measurements (Equation 3). The transmission density was doubled (Equation 4) in order to obtain a reflection density ($d_R$); factors such as trapping, which would lead to increased OD have been ignored, although some allowance has been made for surface scatter. The validity of this approximation has been confirmed by transmission and reflection readings made on thin, uniformly dyed samples.

$$d_T = \text{mass} \times \text{extinction} \quad (3)$$

$$d_R = 2 \times d_T \quad (4)$$

The reflection densities have then been converted (Equation 5) to reflection intensities ($i_R$) in order to obtain (Equation 6) a total for the reflected light ($I_R$) over the whole pixel area. This has then been converted (Equation 7) back to an average reflection density ($D_R$).

$$i_R = 10^{-d_R} \quad (5)$$

$$I_R = \sum i_R \quad (6)$$

$$D_R = -\log_{10} I_R \quad (7)$$

Comparison of modelled and measured values for the OD of prints showed extremely good correspondence over a wide range of print conditions and types of materials once the initial parameters had been set up. It has proved to be an extremely useful tool for predicting the behaviour of new types of materials and has guided the programme of materials development.

For example, Figure 10 shows a comparison of the density build up predicted by the model with two different drive voltages applied to the thermal head. The good correspondence between modelled and measured values then gives confidence to predictions about conditions that are currently inaccessible, so that we can, for example, ask what happens if the head voltage is increased by 50%. The existing head would be destroyed under these conditions, but there is a good possibility that heads with this capability will be available in future. Similarly it is possible to make informed predictions about the properties of materials that are not yet available (such as 1 micron thick base film).

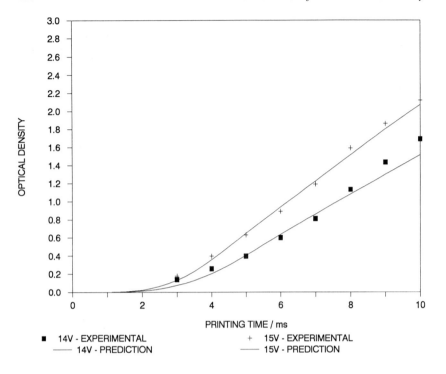

**Figure 10  Comparison of Modelled and Measured Transfer**

## 5 CONCLUSIONS

All printing processes depend on the use of subtractive primary colours to represent the desired image by absorbing light so as to build up the desired shade. In the D2T2 process, this is achieved by transfer of the three subtractive primary colours sequentially from different sections of a colour ribbon onto a receiver paper.

Colour build up in the D2T2 process is achieved by thermal migration of dyes down a concentration gradient. The transfer occurs entirely in condensed phases, and the question of sublimation does not occur for normal print conditions. The modelling process outlined in this paper has been found to give very good predictions of system behaviour under a wide range of conditions, thus justifying its use as a tool, and also confirming the validity of the basic partition/diffusion model.

## 6 ACKNOWLEDGEMENTS

I wish to thank N C Beck for his major contribution to the work, D J Toms for encouragement and helpful discussions, I B Parker and A Jones, J King-Hele, and S D R Watson of Manchester University for assistance with the modelling work, and P W Webb of the Wolfson research unit at Birmingham University for carrying out the thermal measurements.

## 7 REFERENCES

1. See, for example, F. Valenti, 'Thermal Printhead Technology', IGC Thermal Printing Conference, Amsterdam, 22-24 May 1985.
2. See, for example, T. Gotoh, Y Kobori, O Hattori, and K Hanma, **Proc. 3rd Intl Cong. Adv. Non Impact Printing Tech.**, SPSE, San Francisco, August 1986.
3. P. Gregory, **Chem. Brit.**, 1989, 25, 47.
4. R. A. Hann and N. C. Beck, **J. Imaging Technology,** in the press.
5. R. A. Hann and P. W. Webb, **to be published.**

**Materials for Insulation and Optical Data Storage**

New Developments on Isocyanate-based Casting Resins and
Polyurethane Compounds
J. Franke and H. P. Muller (Bayer AG, West Germany)


Materials for Optical Data Storage
M. Emmelius, G. Pawlowski and H. V. Vollmann
Hoechst AG, (West Germany)

# New Developments on Isocyanate-based Casting Resins and Polyurethane Compounds

## J. Franke[1] and H. P. Müller[2]

[1]BAYER AG, KREFELD-UERDINGEN, FRG
[2]BAYER AG, LEVERKUSEN, FRG

**Summary**

Reaction resins based on isocyanates have been in use for the electrical industry for the last 25 years. Especially room temperature-curing, tough resins or flexible systems have found numerous applications. This article presents several new developments which are far superior to conventional resins concerning the compatibility, thermal endurance and tear resistance. Halogen free EPIC resins with inherent flame retardance for heat-class H-applications are discussed as well as the newest results of the 15 years outdoor studies with PUR-insulators. Users of casting resins will have numerous ideas for further applications after reading this article.

**Introduction**

More than 50 years ago Prof. Dr. O. Bayer received the first patent for the reaction of isocyanates with polyols to form macromolecular compounds. Since then the chemistry of polyurethanes experienced an unequalled triumphal advance. The worldwide consumption in 1988 was about 4 million tons. No other plastic material offers such a variety of possible applications as polyurethane does: rigid, flexible and integral foams, elastomers, fibres, lacquers, varnishes, coatings, adhesives and last but not least casting resins.

For 25 years the PUR-components Baygal®, Baymidur® and Blendur® are now applied successfully in the electrical industry. The applications extend from cable jointing over the production of transformers, capacitors and insulators to the encapsulation of small electrical and electronic parts.

Regarding the initial products, there are many different combinations and variations possible enabling a precise adjustment of properties to the requirements. Due to this, and the well-balanced spectrum of properties, polyurethanes are frequently preferred to other raw materials. Another reason for the steady growth of the consumption of polyurethanes as casting resins is the excellent ratio of price to efficiency.

This article reports on progress with new developments made on the range of polyurethane raw materials, which result in an improvement of properties as well as an increase in processing safety.

## State of the art

The approved standard systems for the production of compact, blister- and pore-free insulating material out of PUR consist mainly of:

- isocyanates, mostly technical diisocyanatodiphenylmethane (MDI), which is composed of a mixture of 4,4'- and 2,4'-MDI-isomers and parts of polyisocyanates with 3 or more benzene-cores

- polyols, particularly polyether- and polyesterpolyols

- fillers like silica flour, dolomite, aluminiumhydroxide, chalk

- zeolites.

Corresponding isocyanate reaction resins can be processed according to the usual methods of the cast resin technology [1].

One typical characteristic of these resins is unfavourable: the deficient compatibility of isocyanate with polyol. Contrary to other resins, non-catalysed MDI-polyetherpolyol resins require several minutes of stirring until the compatibility and homogeneity of the initial products is obtained by the formation of urethane-structures (Fig. 1).

Figure 1:

Compatibility of different PUR casting resins
left:  standard PUR after 10 min still incompatible
right: new version, direct compatible after 10 sec

**Direct compatibility**

The incompatibility can be avoided if special MDI-prepolymers are used. By the choice of the right initial isocyanate, the unavoidable increase in viscosity through the prepolymerisation can be limited. The advantage of direct compatibility is not achieved at the expense of the heat deflection temperature.

The comparison of properties is shown in table 1. This direct-compatible isocyanate is in combination with standard polyether systems suitable for cable jointings according to VDE 291 [2]. The increase of safety in processing with this isocyanate is of great advantage. When used in uncatalysed systems, this prepolymer gives a "real" processing time at room-temperature of more than 60 minutes, giving high filled isocyanate-based resins for the first time the chance to be processed with injection moulding machines.

**Table 1:** Comparison of properties of standard PUR casting resins with the direct compatible version

|  |  | standard-system | direct compatible system |
|---|---|---|---|
| viscosity of mixture | mPa·s | 450 | 550 |
| incompatibility | min | 15 | 0 |
| processing time | min | 60 | 60 |
| tensile strength | MPa | 62 | 66 |
| flexural strength | MPa | 110 | 110 |
| Shore D hardness | - | 82 | 88 |
| heat deflection temperature according to Martens | °C | 70 | 70 |

Standard-system: 100 p.b.w. Baymidur K 88
100 p.b.w. Baygal K 55

Direct compatible system: 120 p.b.w. Baymidur VP KU 3-5006
100 p.b.w. Baygal K 55

**Processing with the pressure-gelation-technique**

This efficient processing method once developed for Epoxy resins [3], can now be used with new PUR raw materials for the pressure-gelled production of cured PUR-resin compounds. The two components - polyols, additives, fillers, pigments and zeolite on one side, and the polyisocyanate on the other side - are separatly prepared and degassed. In a two component machinery the resin and hardener were delivered separately and combined in a static or dynamic mixer (Fig. 2).

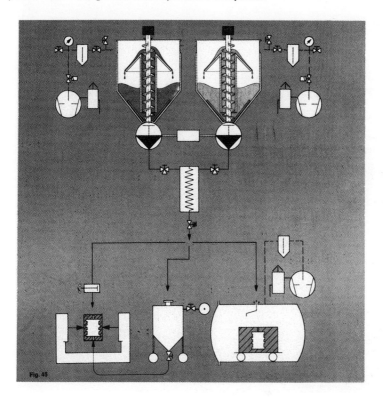

Figure 2

Schematic representation of degassing mixer, flow mixer and vacuum casting or pressure gelation mould equipment


This reactive PUR mass with a temperature of 20 to 50 °C is injected under pressure into the closing-unit. During the filling procedure and the curing, a pressure of 2 to 5 atm is held to compensate the volume shrinkage. The gelation starts at the 180 to 150 °C hot walls of the mould and continues steadily in the direction of the centre. The curing time depends on the size of the part and the temperature used. Actually, a curing time of 2 to 10 minutes is sufficient in most cases. This is much below that of comparable epoxy resins (Fig. 3).

Figure 3: 2-component machine and closing unit for the processing of PUR

**Permanent service temperature**

Characteristic for polyurethanes is their reversion, i.e. the reverse reaction of the urethane bonding into the initial products at higher temperatures. According to IEC 216 [4] soft and flexible polyurethanes have a limiting temperature of 90 °C extrapolated to 20 000 h, whereas tough systems reach 120 °C. Progress is made by the chemical modification of the polyetherpolyols with special epoxy resins. This trick slows down the reversion and prevents it. With this modification the permanent service temperature of flexible systems will be increased to 120 °C and

# Isocyanate-based Casting Resins and Polyurethane Compounds

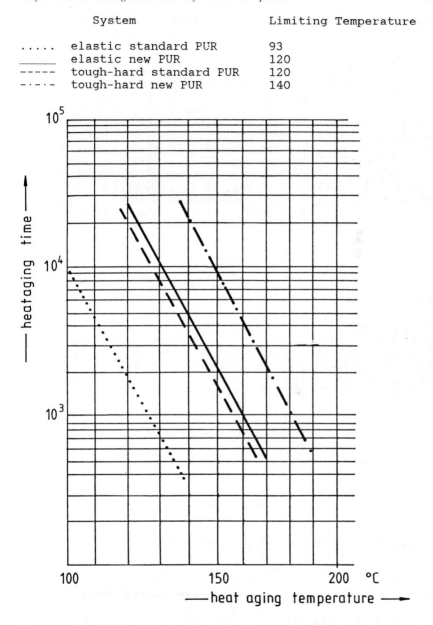

Figure 4: PUR systems of high thermal endurance according to IEC 216

that of tough systems to 140 °C (fig. 4) [5]. By
this, polyurethanes reach heat class B, as hot-curing
epoxy resins do.

## EPIC-systems with superior thermal properties

In comparison to conventional resins with a dimensional heat stability below 150 °C the newly developed EPIC-resins lead to compounds with far better thermal properties. They are able to reach a glass transition temperature of more than 300 °C, a dimensional heat stability exceeding 250 °C and a permanent service temperature of 200 °C according to IEC 216 [6]. These EPIC-systems are based on a combination of isocyanate with epoxy, without the additional use of polyols. The trimerisation of the isocyanate is implemented with special catalysts and the reaction between isocyanate and epoxy results in networks being linked over isocyanurate- and oxazolidinone-groups (fig. 5). The high proportion of heterocycles is responsible for the outstanding thermal properties and the excellent inherent flame retardancy. Table 2 shows a comparison of various thermosetting plastics with EPIC-resins (EPIC = combination of epoxy with isocyanate).

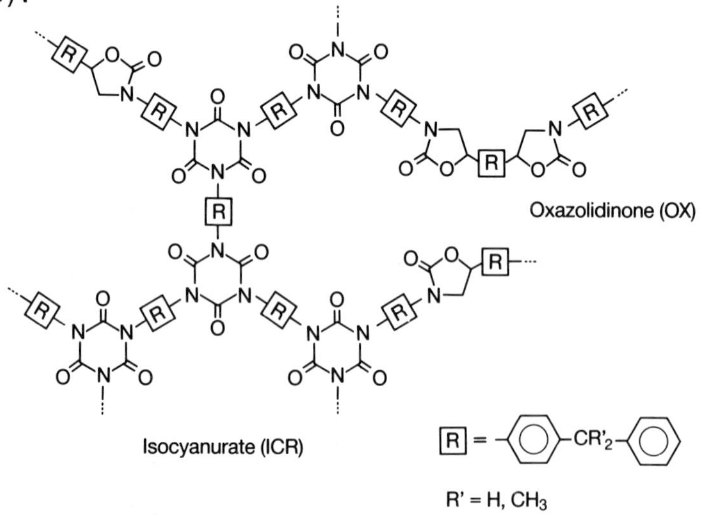

Figure 5: Idealized structure of EPIC compounds

# Isocyanate-based Casting Resins and Polyurethane Compounds

Table 2: Comparison of various unfilled thermosetting plastics (according to DIN 16946)

| | unsaturated Polyester UP | epoxy EP | polyurethane PUR | EPIC |
|---|---|---|---|---|
| form supplied | liquid | liquid or solid | liquid | liquid or solid |
| number of components | resin + catalyst | resin + hardener | resin + hardener + Zeolith | resin + catalyst |
| hardening reaction | polymerisation | polyaddition | polyaddition | polyaddition |
| hardening temperature | rt up to 160 °C | rt up to 180 °C | rt | 180 up to 250 °C |
| compound | hard | hard | flexible up to hard | hard |
| exotherm during curing | very high | high | very low | high |
| **Mechanical properties** | | | | |
| tensile strength [N/mm$^2$] | 50 - 80 | 60 - 80 | 60 - 80 | 50 - 70 |
| elongation at break [%] | 2 - 4 | 3 - 8 | 5 - 50 | 2 - 3 |
| flexural strength [N/mm$^2$] | 100 - 140 | 80 - 140 | 60 - 140 | 100 - 120 |
| impact strength [kJ/m$^2$] | 8 - 15 | ca. 20 | ca. 40 | ca. 10 |
| glass transition temperature [°C] | 70 - 150 | 140 | -40 - 130 | 300 |
| tensile modulus [N/mm$^2$] | 3500 | 4000 | 3000 | 4000 |
| UL 94 flammability at 3.2 mm | failed | failed | failed | V O |
| permanent service temperature IEC 216 [°C] | ca. 110 - 130 | 155 | 90 - 140 | 198 |
| volume resistivity [ohm.cm] | - | $10^{16}$ | $10^{16}$ | $10^{16}$ |
| dielectric dissipation factor | - | 0,005 | 0,006 | 0,009 |
| dielectric constant | - | 3,4 | 3,6 | 3,7 |

The application of this type of reaction, which has been known for many years now, was formerly rendered more difficult by the complicated production and processing conditions. In recent times, substantial improvements have been developed to enable the use of this class of products for many applications.

Mixtures with a good storage stability

Even without the use of catalysts the storage stability of isocyanate-epoxy-blends is limited. The development of a reaction-inhibitor or "stopper" led to a storage stability of the complete mixture for more than 6 months without an influence on the reactivity after the addition of the catalyst. This offers the possibility to present the ready-for-use formulations to the end-user [7].

Dirigible pre-reaction

Of even more importance for the practice is the possibility to start and stop the polyaddition at each point before the gelling in order to receive storage stable products [8]. This gives a range of products with different viscosities up to a solid resin.

Three versions are offered to the market (Table 3): Besides the physical mixture of isocyanate with epoxy (A-state), which is mainly suited as an impregnating and dipping resin, there is a resin of medium viscosity (2000 mPa·s, AB-state) which makes manyfold casting applications possible. The solid resin (B-state) opens up further applications. It is fusible or soluble in organic solvents like acetone or methylethylketone. All these three products are fully compatible with each other. By combining and mixing them the properties of the resin can be adjusted to the respective application. After the curing, the compounds have the same properties independent from the state of polyaddition from which the reaction was started.

Table 3: EPIC-resins; three new ways

| EPIC-Systems | Applications |
| --- | --- |
| A-stage<br>(25 °C) ca. 50 mPa·s | - impregnating resin for rotating machines<br>- dipping resin for small motors |
| AB-stage<br>(25 °C) ca. 2000 mPa·s | - casting resin for electrical and electronic components<br>- laminating resin for filament winding |
| B-stage<br>solid: MP 50 - 60 °C | - transfer moulding compounds for the encapsulation of electronic components<br>- laminating resin for printed circuit boards |

## New catalysts

New types of resins require new catalysts. The systematical examination of the kinetics of the reactions led to the development of a range of catalysts. This gives a choice of different reactivities.

With an amine-based catalyst a pot-life of 4 - 8 h (at room temperature) is obtained. An amine-phosphate-complex [9] as a chlorine-free, latent catalyst remains stable in a resin for approx. 2 weeks but leads at 160 °C to a gelation after a few minutes.

The third and most practical version is a one-component-system including a complex catalyst with a storage stability of more than six months. Heating up this system to 120 °C results in a gelation after a few minutes.

## Properties and Applications

Casting resins based on EPIC reach heat class H. That means, a permanent service temperature above 200 °C over 20000 hours is achieved. Fig. 6 to 8 represent the excellent mechanical and electrical behaviour of EPIC-compounds.

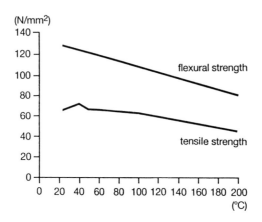

Figure 6: Tensile and flexural strength of EPIC-compounds as a function of temperature

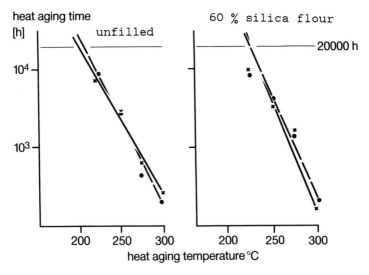

Figure 7: Thermal stability of EPIC-resins

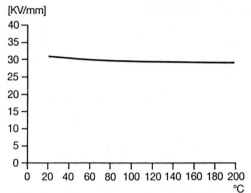

Figure 8: Dielectric strength of EPIC-compounds as a function of temperature

Shown is the flexural and dielectric strength in dependence on the temperature up to 200 °C. The material exhibits in the filled stage a halogen-free flame retardance reaching an UL 94 rate V0-classification. It is clear that new materials possessing high-grade insulation properties hold good prospects for the electrical insulating sector.

A selection of various applications is given in table 3. For example, high performance motors of heat class H have successfully been dipped and impregnated with EPIC and are already in use.

## Cycloaliphatic PUR-compounds

Whereas only new developments and latest findings were treated in this article so far, the next section reports on special applications for which cycloaliphatic PUR-systems have been tested now since 1975: weather resistant compounds for outdoor insulators and transformers [10]. The practical experience of more than 10 years justifies the report in this article.

For out door insulators the use of glass or porcelain is dominating. Since 1960 cycloaliphatic EP-resins are used in 20 kV high voltage circuit lines. How ever, over the time, problems arose concerning surface erosion and destruction of the insulators. That was the consequence of the interference of the insulating material with the environment of oxygen and water under thermal and electrical conditions. The hydrolysis of the ester-bonds in the cured EP-resin was the result and the reason for the decomposition.

The positive experience collected for many years with aliphatic PUR-coatings (DD-coatings) initiated the development of weather resistant PUR-compounds. Especially because of the required light- and hydrolysis-resistance as well as the high tracking resistance, only non-aromatic systems are suitable. Compounds based on isophoronediisocyanate and polyetherpolyol fulfil all the requirements [11]. The absence of hydrolytically unstable ester-groups should give higher life expectancy.

The differences between both systems in short-term laboratory tests as well as in the long-term behaviour in service are pointed out in the following paragraphs.

### Short-term tests according to VDE 0441/1 [12]

To judge whether the compounds are qualified for outdoor use three high voltage tests are chosen for the VDE-standard and the results are classified. The following three tests were measured:

- high voltage tracking resistance (HK)
- high voltage arc resistance (HL) and
- high voltage diffusion dielectric strength (HD)

The results are shown in table 4. The systems having a zero-classification in one of the three tests are not recommendable for outdoor-use. These are the aromatic epoxy and PUR-systems. Whereas cycloaliphatic epoxy resins fulfil the requirements only with silane-treated silica flour, cycloaliphatic polyurethanes are qualified with untreated silica flour and even with a combination of silica flour and aluminium oxide trihydrate, too.

Table 4: Outdoor qualification of reaction resin compounds (Cy: cycloaliphatic, Ar: aromatic); short-term tests according to VDE 0441/1

| Compound system | | Measured value | | | Classification[4] | | |
|---|---|---|---|---|---|---|---|
| | | HK [kV] | HL [s] | HD [kV(min)] | HK | HL | HD |
| Cy-PUR[1] | QM[3] | 4.5 | 216 | 12 (>1) | 3 | 1 | 2 |
| | QM + Silane | 4.5 | 220 | 12 (>1) | 3 | 1 | 2 |
| | QM + ATH | 4.5 | 320 | 12 (>1) | 3 | 3 | 2 |
| Cy-EP[2] | QM | 3.5 | 200 | 6 (<1) | 2 | 1 | 0 |
| | QM + Silane | 3.5 | 205 | 12 (>1) | 2 | 1 | 2 |
| | QM + ATH | 3.5 | 210 | 6 (<1) | 2 | 1 | 0 |
| Ar.-PUR | QM + Silane | 2.5 | 150 | 12 (>1) | 1 | 0 | 2 |
| Ar.-EP | QM + Silane | <2.5 | 185 | 12 (>1) | 0 | 1 | 2 |

1 based on isophoronediisocyanate and polyol
2 based on hexahydrophthalic acid diglycidylester and hexahydrophthalic acid anhydride
3 QM: silica flour; ATH: aluminium oxide trihydrate
4 3: very good; 2: good; 1: suitable; 0: unsuitable

## Long-term practical experience

In cooperation of Bayer AG with Siemens AG and Schleswag practical tests were carried out with 20 kV hanging insulators on the isle of Nordstrand and with 20 kV pin insulators in Bergheim/ Erft. The insulators have been in use for 8 and 13 years respectively without any failure or breakdown. Merely the surface of the insulators shows obvious signs of ageing. While on PUR-insulators only a plain erosion occurs (that means a surface roughness and a loss of gloss in comparison to new insulators), epoxy insulators show a "pitting" erosion. These results are illustrated in figures 9, 10 and 11.

Figure 9 number of insulators with pitting erosion in dependance on time

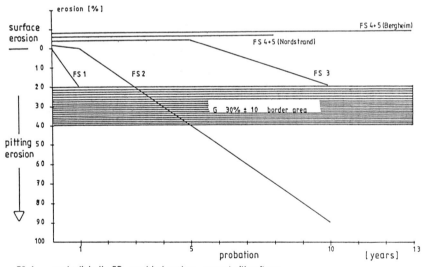

FS 1: cycloaliphatic EP- moulded system, normal silica flour
FS 2+3: cycloaliphatic EP- moulded system, silanized silica flour
FS 4 +5: cycloaliphatic EP- moulded system, normal and silanized silica flour

Figure 10: Insulator made of cycloaliphatic epoxy (filler: Silane-treated silica flour) after 10 years of outdoor-use, the pitting erosion is obvious

Figure 11: Insulator made of cycloaliphatic PUR, all specimens are unobjectionable, no erosion. a,b: filler: usual silica flour; 8 years of outdoor-use; c: new insulator; d,e: filler: silane-treated silica flour: 8 years of outdoor-use

## Service life prognosis according to IEC 216

For the forecast of the duration of life of those insulators Näcke et al take the high voltage diffusion dielectric strength (HD) after 49 days of storage in boiling water. As illustrated in fig. 9 the HD of cycloaliphatic PUR specimens decreases asymptotically with increasing time and stabilizes itself at about 10 kV/cm whereas the HD of cycloaliphatic EP runs towards zero. Here the critical boundary line between a disruptive discharge through air and one through the insulating material is already passed after 1 day. The changes in the appearance of the samples were significant. The analogy to the figures 4 and 5 of the used insulators is evident [13]. Thus the practical tests confirmed the laboratory tests very well.

Conclusion

There is a fundamental difference between the two cycloaliphatic materials regarding the thermic-hydrolytical behaviour. Epoxy compounds are irreversibly, autocatalytically saponified upon diffusion of water into the material because of the great proportion of ester-groups. The cycloaliphatic PUR is distinguished by the thermic-hydrolytical stability because of its urethane and ether groups which are more resistant against hydrolysis. Only a physically reversible interaction takes place under outdoor conditions. Whereas a pitting erosion on the EP-insulators takes place, the PUR-insulators are satisfactory after 10 years of practical use. By the correlation of laboratory results with practice results and extrapolation there are more than 25 years of sevice life to be expected. In the meantime not only insulators are in use but also outdoor-transformers (made successfully by PUR-coating resins).

Literature

1   F. Ehrhard: Gießharze in G. Oestel Kunststoffhandbuch, Band 7, Polyurethane; Carl Hanser, München, New York, 1983

2   VDE 0291: Casting resins for cable jointings, Beuth, Berlin, 1979

3   R. Kubens, H.-D. Martin: Kontakte Studium, Band 6, Grafenau, Lexika-Verlag, 1976

4   IEC 216: Guide for the determination of thermal endurance properties of electrical insulating materials, Beuth, Berlin, 1980

5   DE-Patent 3340588, Bayer AG

6   R. Kubens, Kunststoffe 69 (1979), 8, 455

7   DE-Patent 3807660, Bayer AG

8   DE-Patent 3600764, Bayer AG

9   DE-Patent 3644382, Bayer AG

10  E. Ehrhard: Kunststoffe 74 (1984), 2, 99

11  DE-Patent 2338185, Bayer AG

12  DIN-VDE 0441, Testing of materials for outdoor insulators, Beuth, Berlin, 1985

# Materials for Optical Data Storage

M. Emmelius, G. Pawlowski, and H. W. Vollmann

HOECHST AKTIENGESELLSCHAFT, GESCHÄFTSBEREICH
INFORMATIONSTECHNIK, WERK KALLE-ALBERT, D-6200 WIESBADEN, FRG

INTRODUCTION

The mass storage units of the next generations of computers will be based on optical processes having a storage density which exceeds that of all hitherto known storage techniques. Since 1982 read-only memories in the form of compact discs (CD-ROM; compact disc - read only memory) have become very successful in the field of audio electronics. Research and development are now concentrated on materials for write-once (WORM; write once read many) and rewritable (EDRAW; erasable direct read after write) storage systems.

Suitable materials for optical data storage are substances in which data markings can be recorded and deleted respectively using semiconductor lasers. The development is centered on the synthesis of infrared-absorbing dyes for WORM memories and the production of rare earth/transition metal alloys for magneto-optical (EDRAW) data recording.

Optical Memories

Research has been in progress on optical memories since the end of the sixties. However, it was not until 1982 that the breakthrough finally came with the compact disc. Audio and video CDs are optical memories which are based on a principle similar to that of gramophone records: information items are encoded in a structural pattern, from which the original information can be regained by means of a playback unit. In contrast to gramophone records, the readout takes place by a noncontact method using laser light.

The information is impressed into a plastic plate and cannot be altered. The information carriers are pits which are impressed into the surface of a plastic disk and are separated by the relief positions, the land. The pits, which are arranged in sequence and which have a depth of approximately 0.5 µm and a width of 0.8 µm, form tracks at a spacing of approximately 1.6 µm. The disk consists of an injection-molded polycarbonate or polymethyl methacrylate substrate (diameter 12 cm) and is coated on the information-carrier side with a highly reflective aluminium film, which is protected against mechanical damage by a polymer layer. The light of the playback unit reflected by the aluminium mirror is passed to a photodetector. When a pit is swept over, the signal intensity of the laser beam decreases significantly since the light is scattered by the pit structures. In this way, it is possible to read binary data which are encoded in accordance with specified algorithms: reflection results in a high signal intensity and is recorded for example as 1; a signal intensity which is reduced by scattering represents 0.

The compact disc system developed by Philips and Sony has meanwhile become a world standard, and has been licensed to many other companies. There are as yet no standardized formats for the WORM and EDRAW data disk. As far as the data disks are concerned, the main difference from the CD format is that the substrates are provided with one or more guide tracks which, using a laser beam, permit orientation on the initially uninscribed disk. The impressing of the guide track, also referred to as a groove or track, creates elevated regions to receive the data markings.

Write-once optical memories are designated as WORM systems. In contrast to the CD-ROM, it is possible for the user to record information. The principle of all WORM memories is based on an irreversible physical transformation of the recording medium, which is initiated by the irradiation of laser light. The following processes have been described: formation of pits, bubbles, bumps, changes to the surface texture, phase change, switching of liquid-crystal phases, alloying of metal films and photochromic color change. In most cases, the reflectivity of the recording medium is altered by the marking process. This takes place locally at positions heated by the laser beam so that data spots are created, which are distinguished from the surrounding medium by their reflectivity.

The physical marking process of EDRAW media is reversible, and thus enables the user to erase data which have been inscribed. EDRAW memories have been developed on the basis of magneto-optical materials, reversible phase transformations and photochromic dyes. In addition, there exists a series of other reversible storage processes to which no technical significance is yet, however, ascribed.

Production of Substrates

The production is divided into the fabrication of the master and the preparation of substrates. Injection-molding dies are cast in a multistage process (mastering) which involves the latest technologies of microelectronics.

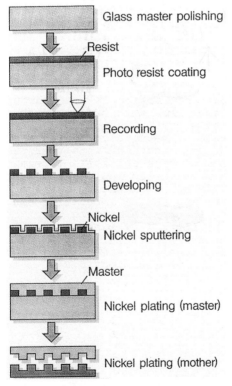

Fig. 1 Mastering process of the stamper for the injection molding optical disk.

The mastering process begins with the cleaning and polishing of a glass disk which is required to construct the master. An adhering layer is applied by vapor deposition using an agent creating water-repellent properties, e.g. hexamethyldisilazane (HMDS), which coats the surface with a positive photoresist. The surface is exposed to a UV argon ion laser, which inscribes the pit structure in accordance with the digital information as played back from the master magnetic tape. During subsequent development the exposed resist is dissolved away. The photoresist and the glass surface are then coated with a layer of metal (nickel in most cases) approximately 50 – 100 nm, either by vacuum vapor deposition or by sputtering. Further metallic layering to form the actual metal master is carried out using a currentless or electrolytic process. After separating the master from the glass base, parent masters are produced by electroplating, and the individual injection-molding masters are formed from these parent masters by means of a second electroplating operation. By setting up such a "family of masters", it is possible to prepare new ones immediately without repeating the costly mastering process.

In most cases, the information is read through the disk, i.e. the laser beam must pass through the substrate material without loss of intensity and – in the particular case of magneto-optical systems – without change of polarization. This requires a substrate material of high optical homogeneity, purity and optical transmission and low birefringence. Bubbles, gel particles, inclusions, surface defects, and zones of turbidity due to impurities can give rise to scattering and absorption effects which cannot be compensated by the servo functions of the electronic and optical systems. Significant properties of the disk materials with regard to mechanics, optics, environment and productivity are represented in Fig. 2 for polycarbonate (PC), polymethyl methacrylate (PMMA).

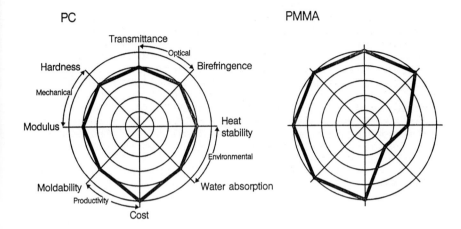

Fig. 2. Properties of polymeric substrate materials. Parameters are represented regarding optical, mechanical and environmental properties and productivity of polycarbonate, polymethyl methacrylate. The performance of a polymer improves from the inner to the outer circle.

A further criterion for the selection of a substrate material is its absorption of water, since swelling and thus distortion of the disks can arise as a result of water vapor. The sensitivity to moisture of certain recording media is very great, so that the water absorption of the substrate may be the limiting factor with respect to its use.

The birefrigence of the flat polymer disks must be as small as possible; for example, it is specified at ± 50 nm for the Compact Disc. The process of molding storage disks must accordingly be adapted to the specific birefringence properties of the polymer; in addition to this, it is necessary to include the effects of other production steps. Reduction of the birefringence of substrate disks may be achieved by the use of polymer blends which consist of polymers having different anisotropic polarizabilities.

New types of polymers such as cyclic polyolefins, compounds with very low water absorption, are now investigated for the production of substrates.

Write Once Storage Materials

The initial investigations on write-once materials based on dyes were carried out using non-directly-modulable gas lasers, for example argon ion lasers emitting at 488 nm, owing to the lack of NIR-sensitive dyes. Even at that time, the benefit of this technology became known and was protected under patent law. The dyes which were used included azo dyes, ketocoumarins, quinones, fluorescein, polyester yellow, or diethoxy-thioindigo. Phthalocyanines, naphthoquinones, simple squarylium and cyanine dyes or triarylmethane dyes proved to be suitable for recording information by means of helium-neon lasers (633 nm). The practical significance of these storage media is slight. Nevertheless, it should not be overlooked that it is possible to achieve a reduction of the pit geometry and thus, in principle, a higher storage density of the disk by light of shorter wavelength.

The number of NIR-sensitive dyes has greatly increased during recent years. In practical terms, four categories of dyes have, in particular, achieved significance: 1. methine dyes, 2. phthalocyanine derivatives, 3. specific condensed arenes, and 4. metal complexes. There are also many other examples of dyes which do not fall into this subclassification; some of these will be outlined after the discussion of the four classes of substances.

Methine Derivatives. The first compounds of this category to be investigated were heptamethine cyanines, with benzoxazole, benzothiazole, (benzoannelated) indole or quinoline moieties as terminal groups. In addition to having high absorption coefficients, these compounds exhibit excellent reflection properties (up to 40 % at 800 nm) and thus yield materials with good SNR values and high signal contrast. Indole derivatives are particularly preferred on account of their favorable solubility properties.

The following properties distinguish the methine dyes as absorbers for write-once disks: they have a high molar extinction coefficient and require only small energies to form sharp and intense pits (approximately 0.5 nJ/pit), and a favorable signal-to-noise ratio results from the high reflectivity (> 30 %). Methine dyes are photooxidized by singlet oxygen in the presence of light.

An increase in the stability may be achieved by the introduction of electron-withdrawing substituents, for example halogens or cyano groups, as a result of which a slight bathochromic shift of the absorption maximum takes place in consequence of a reduction of the LUMO (Lowest Unoccupied Molecular Orbital). There are other possibilities for stabilization by introduction of a cyclic unit (Fig. 3) into the methine group. It has been shown that particularly suitable compounds are those derivatives which contain squarylium or croconium groups like those shown in Figure 3 (bottom right) in the methine chain.

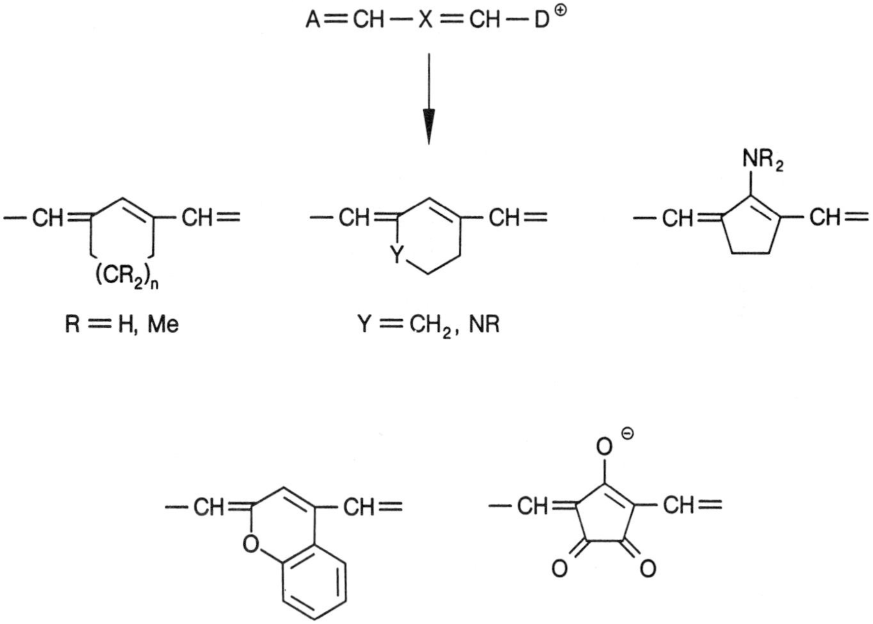

$$A=CH-X=CH-D^{\oplus}$$

$\downarrow$

−CH=⟨(CR$_2$)$_n$⟩−CH=  
R = H, Me

−CH=⟨Y⟩−CH=  
Y = CH$_2$, NR

−CH=⟨NR$_2$⟩−CH=

−CH=⟨O⟩−CH=

−CH=⟨O$^{\ominus}$, O, O⟩−CH=

Fig. 3. Stabilization of methine dyes sensitive to oxidation by introduction of cyclic moieties. A = acceptor, D = donor

These have been combined with the most widely varying terminal groups such as azulenium, pyrylium, benzo or naphthothiapyrylium, naphtholactam or oxoindolizine heterocycles. Furthermore, storage media consisting of polymeric cyanine derivatives or of tetra-(dialkylaminophenyl) cyanines have been proposed. The latter exhibit good solubility properties, but are inferior to the cyanines in their optical properties. However, they are more stable towards singlet oxygen, and the addition of radical traps increases their stability.

The sensitivity to oxidation of the above indicated methine dyes may also be reduced by the addition of an oxygen trap. Examples of suitable compounds are dithiolato complexes of platinum, cobalt and nickel or aminium derivatives, which can be obtained as salts. They are added to the mixture or directly attached to the dye. Alternatively, salts with a cyanine dye as a cation and a radical trap as an anion can be used, e.g. a dithiolato derivative. These salts tend to be highly stable.

A particularly intensively investigated group of methine dye derivatives comprises the squarylium dyes, which have been developed by Philips and Kodak. The compound SQS (Fig. 4) which forms an inner salt can be

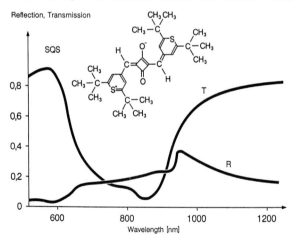

Fig. 4. Transmission (T) and reflection (R) spectra of a 90 nm thick film of the squarylium dye SQS suitable for a WORM medium.
X = relative transmission or reflection

spincoated as a dye solution (approximately 50 g/L n-propanol), on account of the tert-butyl groups which increase the solubility. At 800 nm (Kr ion laser), an SQS film having a thickness of 90 nm reflects approximately 20 %, and transmits approximately 10 % (Fig. 4). Transmission and reflection are functions of the layer thickness; transmission decreases exponentially with layer thickness, while reflections are at a maximum at approximately 90 nm layer thickness in the case of SQS dye. The signal-to-noise ratio is a function of the irradiated energy.

Phthalocyanine Derivatives. Phthalocyanine ($H_2Pc$), a tetraaza [18] annulene or tetrabenzotetraazaporphyrin, has proven to be a useful dye for write-once optical storage disks. In $Pc^{2-}$, four isoindole units are cyclically connected to one another by nitrogen bridges. The two protons of the metal-free substance may be replaced by numerous metal atoms (MPc) with valency three or higher. In this case axial residues situated perpendicularly to the Pc plane can be introduced leading to a considerable improvement in the solubility of the virtually insoluble original substance (Fig. 5).

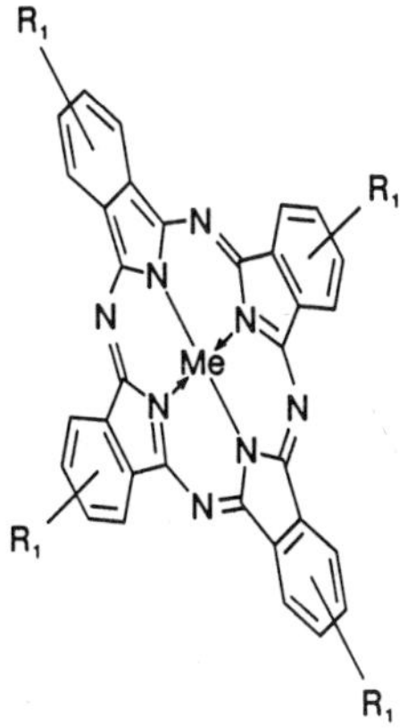

$\lambda_{max}$ = 600 - 800 nm

Me = $Cu^{2+}$, $VO^{2+}$, $Al(OAlkyl)^{2+}$, $Si(OAlkyl)_2^{2+}$, $Ge(OAlkyl)_2^{2+}$

$R_1$ = H, $F_4$, -OAlkyl, -$CH_2$O-Alkyl

Fig. 5. Peripherally substituted phthalocyanine derivatives for WORM media.

In the presence of peripheral substituents, the solubility in aqueous solutions or organic solvents is increased. Unsubstituted phthalocyanines (Pc) are easily synthesized in high yield and are compounds of high chemical stability. More important, phthalocyanines are attractive as optical recording media: they have high absorption coefficients $\epsilon > 10^5$, they are thermally stable, they exhibit virtually no toxicity, they show an excellent stability towards atmospheric effects and they may be applied to the substrate by sublimation or, with appropriate substitution, by spin-coating. Their absorption properties are determined by the rigid, internal $18\pi$ electron system as a chromophore. They absorb especially well in the range of HeNe laser radiation (633 nm). In 1981, Kivits et al. were the first to propose the use of phthalocyanines as absorbers in storage disks of the WORM type and to expand the range of spectral sensitivity of the particularly suitable vanadylphthalocyanine (VOPc), the absorption maximum of which is at 730 nm, by thermal treatment; this results in a broadening of the long-wavelength absorption band beyond 800 nm. Using this material, it became possible to produce a recording medium with a weak SNR with the use of a krypton laser (799 nm). A bathochromic shift and broadening of the longest-wavelength absorption maximum of phthalocyanines, e.g. in the case of MgPc, AlClPc, InClPc or TiClPc, may also be achieved by a THF or $CH_2Cl_2$ solvent or vapor treatment.

It is possible to achieve a bathochromic shift of the longest-wavelength absorption band by the use both of strongly electron-donating and -withdrawing substituents. Thus, in particular, hexadecafluoro derivatives of $Cu^{2+}$, $Zn^{2+}$, $Pb^{2+}$ or $SnCl_2^{2+}$ show high absorptions above 800 nm and are suitable for irradiation with semiconductor lasers. However, if the conjugated system is enlarged by benzoannelation to give the naphthalocyanines (Nc) (Fig. 6), then a substantial bathochromic shift of the long-wavelength absorption band may result. While the absorptions of the 1,2-naphthalic acid derivatives show, in this case, a slight bathochromic shift by 20 nm, in the case of the 2,3-naphthalic acid derivatives shown in Fig. 6 this effect is significantly more marked: displacements by more than 10 nm to 752 (PdNc) to 855 (MnNc($OCOCH_3$)) have been observed. However, by enlarging the $\pi$-system, the derivatives are less thermally stable and considerably difficult to synthesize.

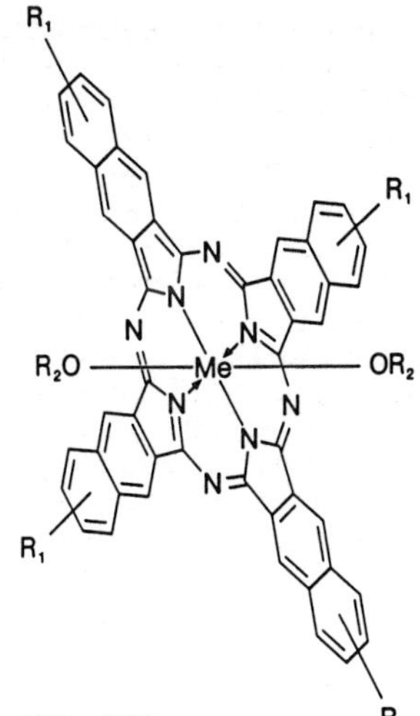

$\lambda_{max}$ = 720 - 820 μm
Me = Si, Ge
$R_1$ = H, t-$C_4H_9$, -O($CH_2$-$CH_2$)$_n$-OAlkyl, -COOAlkyl
$R_2$ = -Si(Alkyl)$_3$, -Si(Alkyl)$_2$-O-Alkyl

Fig. 6. Peripherally and axially substituted naphthalocyanine derivatives for WORM media.

The biaxial derivatives of silicon-naphthalocyanine (SiNc$^{2+}$) (Fig. 7) have been noted for their economical synthetic properties as well as their outstanding optical properties.

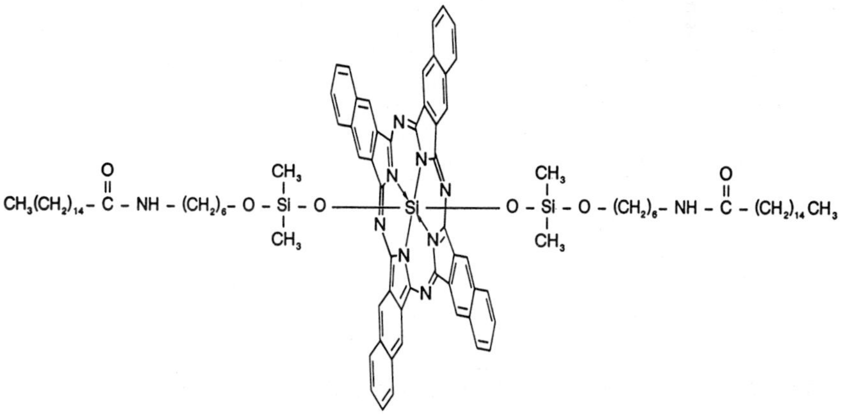

Fig. 7. Bisaxially substituted silicon-naphthalocyanine for WORM media.

The axial substituents can be selected so that they guarantee adequate solubility as well as representing a type of polymeric matrix preventing the tendency toward crystallization or reagglomeration. Compounds of this type exhibit high absorptions in the range of 780 – 830 nm, a reflectivity exceeding 30 % and a signal-to-noise ratio of > 50 dB. The use of a soluble, axially substituted aluminium-naphthalocyanine has been reported; in this case, at a write wavelength of 830 nm a sensitivity of 0.7 nJ/per pit, a reflectivity of 25 %, a signal-to-noise ratio of 55 dB and a reduction of the signal-to noise ratio after $1.5 \times 10^6$ reading cycles by less than 1 dB (1 mW readout energy at 830 nm). The most recent patent applications claim naphthalocyanine derivatives for WORM memories, which are both axially and also peripherally substituted.

The naphthalocyanine derivatives represent an interesting alternative to methine dyes as absorbers of diode laser radiation. Initial prototypes of WORM disks based on naphthalocyanines have already been put on the market.

Quinoid Polynuclear Aromatic Compounds. A further category of dyes to which reference is frequently made in the literature and which can be used as an absorber layer for WORM storage disks comprises the quinoid polynuclear aromatic compounds. The use of these compounds is based on a different concept: the dye molecules are rather small, generally of low solubility and, in contrast to the methine dyes, uncharged, so that they may be applied to the substrate by means of vapor deposition techniques (Figs. 8 and 9).

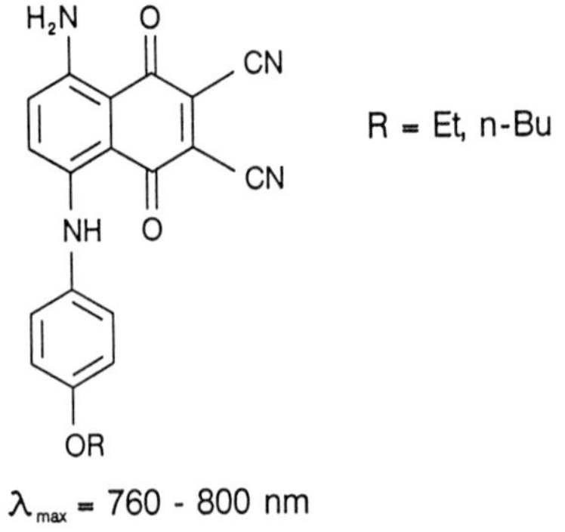

R = Et, n-Bu

$\lambda_{max}$ = 760 - 800 nm

Fig. 8. 5,8-Diamino-1,4-naphthoquinones absorbing in the near infrared.

The most widely investigated group of dyes are amino derivatives of the 1,4-naphthoquinone type (Fig. 8), which have already been known for a long time in the dyeing industry as pigments with a broad color spectrum. In terms of their reflectivity and sensitivity properties, they are less attractive than the known cyanine dye systems but offer higher chemical stability and give storage media with long service life of the data. The systematic examination of 1,4-naphthoquinone derivatives absorbing in the NIR range was extended by Matsuoka et al. to anthraquinone, phenothiazinequinone and phenoselenazinequinone derivatives (Fig. 9) and previously developed synthetic strategies by means of PPP molecular orbital computations. In this case, it became evident that the long-wavelength absorption maximum was influenced both by a change in the donor- and in the acceptor-moieties, since the coloration of these compounds is generated by an intramolecular charge transfer. Derivatives of the type shown in Fig. 8 contain the combination of a suitable donor and strong acceptors. In solution, they absorb at approximately 770 nm, with high molecular extinction coefficients, and respond to the radiation of commercial laser diodes.

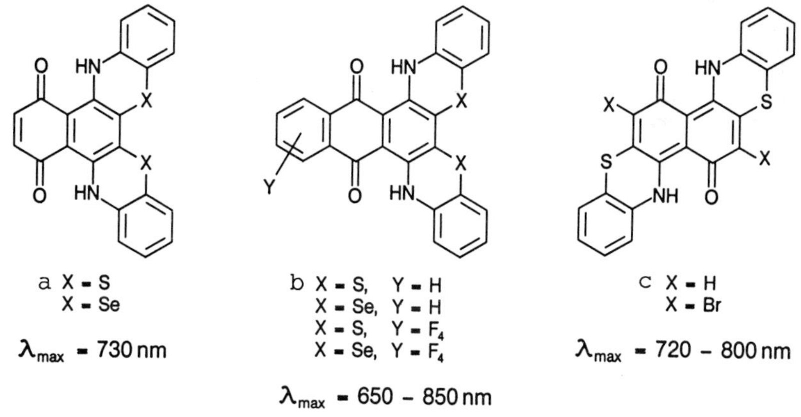

a  X = S
   X = Se

$\lambda_{max}$ = 730 nm

b  X = S,  Y = H
   X = Se, Y = H
   X = S,  Y = $F_4$
   X = Se, Y = $F_4$

$\lambda_{max}$ = 650 – 850 nm

c  X = H
   X = Br

$\lambda_{max}$ = 720 – 800 nm

Fig. 9. Infrared absorbing phenothiazine and phenoseleanzine derivatives with benzoquinone (a), naphthoquinone (b) and 1,4-diketo units (c).

At a signal-to-noise ratio > 50 dB, an acceptable reflectivity of up to 25 % was observed at 830 nm. The substitution of the free amino group by alkyl or aryl radicals leads to a further bathochromic shift. In 1985, NEC announced a WORM storage disk based on the ethyl ether (Fig. 8, R = $C_2H_5$). Although these ethers exhibit optical properties which are on the whole acceptable, they have the disadvantage of high output losses on vacuum coating.

An improvement in the chemical and thermal stability (and thus in the sublimation properties) has been achieved by the phenothiazine or selenazinequinone derivates (Fig. 9) and their isomers. These compounds form homogeneous films with smooth surfaces. Their absorption range is between 750 and 850 nm, and strongly electron-withdrawing substituents have a strong bathochromic effect. Structurally related compounds used include specific indanthrene derivatives, violanthrones, and amino-substituted phthaloyl-acridones.

Metal Complexes. While the phthalocyanine ligand ($Pc^{2-}$) is itself a chromophore, there is a series of colorless, in most cases sulfur-containing ligands which absorb in the visible or near infrared range only as a result of 2 : 1 complexing by means of a metal ion. The long-wavelength absorption occurs due to the formation of a 10 π electron system in the complex. A typical example is represented by the dithiolatonickel complexes (Fig. 10), which can be isolated as salts or neutral complexes and exhibit absorption bands between 715 and 1300 nm in solution.

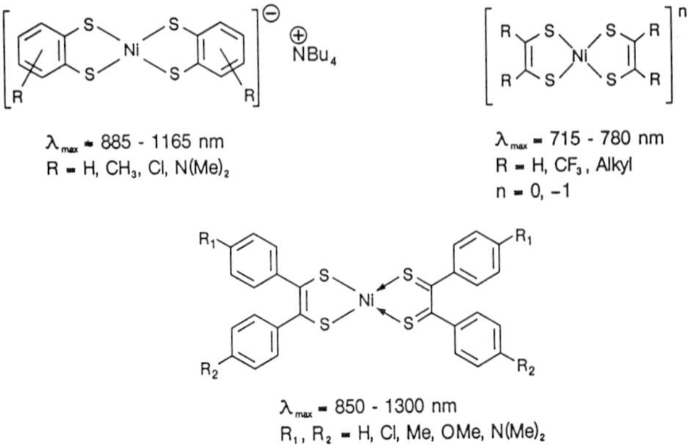

$\lambda_{max}$ = 885 - 1165 nm
R = H, $CH_3$, Cl, $N(Me)_2$

$\lambda_{max}$ = 715 - 780 nm
R = H, $CF_3$, Alkyl
n = 0, -1

$\lambda_{max}$ = 850 - 1300 nm
$R_1$, $R_2$ = H, Cl, Me, OMe, $N(Me)_2$

Fig. 10. Infrared absorption of dithiolatonickel complexes.

Other suitable metals are palladium, platinum or alternatively cobalt, copper, zinc and cadmium.

In connection with the cyanines, the dithiolato complexes are very efficient oxygen traps and thus considerably increase the stability of cyanine absorber layers. At the present time, there is no precise concept concerning the protection mechanism. Closely related classes of compounds include α, α-diimino-cis-1,2-ethylenedithiolatonickel derivates or complexes in which aromatic o-diamines, o-aminothiols or o-selenols ($\lambda_{max}$ = 780 - 900 nm), indoanilines or quinol-inediones are used as ligands.

Other Dyes. In addition to the most important classes of compounds which have been outlined before the patent literature also discloses a whole series of other NIR-absorbing compounds. These include specific aminoarylpentafulvenedicarbonitriles ($\lambda_{max}$ = 680 - 785 nm), ethynevinyl di- and triphenylcyclopropanes ($\lambda_{max}$ ≈ 770 nm), which can be applied by sublimation or, with appropriate substitution, by spin-coating, polypyrrole (1 nJ/pit at 830 nm) and intermolecular charge transfer complexes with 5-nitro-2,3-dicyanonaphthoquinone as acceptor ($\lambda_{max}$ < 740 nm).

## Materials for Reversible Data Storage

Several different materials may be used for reversible data storage, but only a few are at the stage of industrial application. It is required that the information can be easily read and erased, and this leads to a conflict which can be resolved only in circumstances in which a material at room temperature exhibits two conditions which can be selectively addressed and switched. However, many materials which are in principle appropriate are altered in their condition of marking when a laser beam is frequently used for readout. The development of reversibly operating magneto-optical memories and phase transformation media has progressed to a stage at which marketing is imminent.

The Magneto-Optical Effect. The magneto-optical (M/O) effect can be used for the readout of data, since a rotation of the plane of polarization takes place as a result of the interaction of a polarized light beam with magnetic material (Fig. 11). In this way, it is possible to read data markings in reflection (M/O Kerr

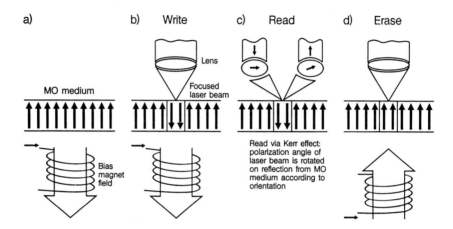

Fig. 11. Principle of magneto-optic (MO) data storage:
a) the perpendicular magnetization of the MO-medium is unaffected by magnetic field at room temperature;
b) recording via heating by laser beam and turnover of the magnetization via a magnetic field;
c) reading via Kerr effect;
d) erasing via heating by laser beam and turnover of the magnetization according to original direction.

effect) or in transmission (M/O Faraday effect). Writing and erasure by means of a laser beam is based on the temperature dependency of the magnetization (Fig. 11). A recording medium magnetized perpendicular to the disk is heated at definite positions, and the magnetization is flipped over into the opposite direction by a magnetic field. Categories of recording materials investigated for technical applications are amorphous rare earth/transition metal alloys (RETM), and ferrites.

Ternary and quaternary alloys of rare earths and transition metals open up a multiplicity of design possibilities for structure/property relations. The search for technically suitable M/O materials extends on a preferential basis to alloys of such transition elements as iron and cobalt with the heavy lanthanoids

*Materials for Optical Data Storage* 199

gadolinium, terbium, dysprosium, holmium, erbium, neodymium, praseodymium and samarium. In addition, films of the most widely varying ternary and quaternary alloys have been described: with cerium, promethium and europium as rare earth elements, and titanium, vanadium, chromium, manganese, nickel and copper as transition metals, as well as various doping elements such as bismuth, tin, lead, germanium, molybdenum, gold, silver, palladium, platinum and uranium.

In contrast to the amorphous RETM alloys, ferrites are air-stable M/O materials which are capable of being used without costly protective layers. Recently, these oxides gained renewed importance after having lost significance for reasons associated with the difficult process technology and inadequate quality of reproduction. Ferrites are mixed oxides consisting of trivalent iron and various metal cations which crystallize with spinel, garnet or hexagonal type structures.

Materials for Reversible Phase Change. Reversible phase change of inorganic materials is the second alternative to the magnet technology which is currently under intensive investigation. Transitions between light-scattering crystalline and highly reflective amorphous phases permit the switching. The crystalline materials - alloys of elements of the fifth and sixth main group - change into an amorphous phase after pulsed laser irradiation from the melt with rapid cooling. The crystalline phase is formed again by extended heating beyond the glass point. The advantages of this storage technique are the simple optical system of the disk drive and high stability of the medium. Disadvantages are the incomplete recrystallization in the event of frequent overwriting, thus deforming the atomic lattice and destabilizing the inscribed information on repeated reading. This can be reduced by matching the laser energy to the number of cycles.

Summary and Outlook.

As a result of the success of CD technology, the development of write-once (WORM) and reversible (EDRAW) storage disks to the stage of marketing readiness has become a matter of prime importance in industry during recent years. The IR emission of the semiconductor lasers employed as light source makes the synthesis of novel IR-absorbing dyes necessary for WORM recording media. Specific compounds of the category of the polymethines, phthalo- and naphthalocyanines and dithiolatonickel complexes have already been used with success.

The most highly developed EDRAW memories are represented by the magneto-optical films consisting of rare earth/element/transition metal alloys. It is expected that they will be able to replace the currently dominant mass memories on the basis of magnet technology within broad fields. Their advantages are high storage density, insensitivity to mechanical stress, and easy exchangeability and thus they follow the development trends of computer technology.

Developments of new optical disk drives will be a great incentive with regard to storage media. The use of optical reading systems based on laser arrays, of integrated optical systems together with polymeric optical waveguides, and of lasers with frequency doubling to generate smaller wavelengths will open up possibilities for further increasing the data rates and storage densities. In the near future, this will also initiate the production of novel recording media. In the far future data storage could achieved possibly on a molecular level (Fig. 12) as it is realized in nature.

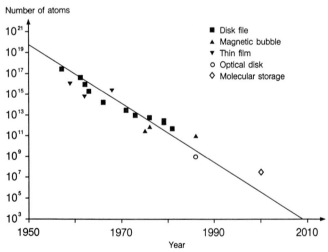

Fig. 12. Comparison for different storage technologies of number of atoms used to store a bit; the line represents the development of future technologies.

**Liquid Crystal Polymers and Nonlinear Optical Materials**

Liquid Crystal Polymers
G. W. Gray (University of Hull)

Polymers for Nonlinear and Electro-optic Applications
R. N. DeMartino (Hoechst Celanese, USA)

Organic Materials for Nonlinear Optical Applications
D. Broussoux, P. LeBarny, J. P. Pochelle and P. Robin
(Thomas-CSF, France)

# Liquid Crystal Polymers

## G. W. Gray
SCHOOL OF CHEMISTRY, THE UNIVERSITY, HULL HU6 7RX, UK

### 1 INTRODUCTION

Liquid crystal polymers are historically fairly new, but the challenge of combining the anisotropic characteristics of low molar mass liquid crystal materials and the rheological properties of polymers in one material, and the implications of this for technological applications have fired activity on the research front and led to a proliferation of books[1-4] reviews[5,6] and papers on the subject.

Liquid crystal polymers (LCP) are macromolecules made up of repeat units which incorporate moieties known as mesogenic units, i.e. components whose molecular structures and shapes are consistent with those commonly associated with low molar mass (LMM) materials which form classical liquid crystal (LC) phases or mesophases.

LMM LC systems involve molecules of three types:

Rod shaped molecules forming Calamitic LC
Disc shaped molecules forming Discotic LC and
Amphiphilic molecules forming Amphiphilic LC

Amphiphilic molecules in their simplest form consist of a polar hydrophilic head group and a hydrophobic tail.

An understanding of LCP systems demands a knowledge[7-9] of the macroscopic ordering that occurs in LMM LC systems, but only a brief synopsis of this can be given here.

<u>Order in Low Molar Mass Liquid Crystal Systems</u>

<u>Calamitic Systems</u>. When the rod-like molecules are self-assembled in a statistically parallel alignment, but

without positional ordering of the molecular centres, the
fluid, anisotropic phase so formed is called a NEMATIC LC.
If in addition to the parallel alignment, the molecular
centres of gravity are distributed in density planes, the
lamellar LC phase is called SMECTIC, and many possibilities
arise for polymorphism of such phases (different tilt angles
to the layer plane, different packing arrangements, etc.).
Finally, if the rods are chiral, helical structures may
develop – e.g., chiral nematic (N*) phases with thermochromic characteristics and chiral smectic C phases ($S_C$*)
that may exhibit ferroelectric and pyroelectric properties.
    <u>Discotic Systems</u>. Here the disc shaped molecules, e.g.
hexa-esters of hexahydroxybenzene, may self-assemble without
ordering of their centres, but with their planes parallel
giving a nematic discotic ($N_D$) LC phase or they may stack up
in columns forming a variety of polymorphic columnar D
phases. $N_D$* phases are also known.
    <u>Amphiphilic Systems</u>. Amphiphilic molecules may self-
assemble together with solvent to form lamellar phases or
phases resulting from an aggregation of micelles such as
discs, rods, spheres or columns – e.g., hexagonal phases
formed by a packing of columns with incorporated solvent.

    Whereas for amphiphilic systems, change in order and
phase type may be effected by increasing or decreasing the
amount of solvent and/or temperature (Lyotropic LC systems),
calamitic and discotic systems undergo change in order and
phase type simply with change in temperature (Thermotropic
LC systems – no solvent involved).

## 2 TYPES OF LIQUID CRYSTAL POLYMER

Two types of LC polymer should first be considered. First
there is the MAIN CHAIN type in which the mesogenic units
are linked end to end, and second the SIDE CHAIN or COMB
type in which the mesogenic groups are appended to a flexible backbone. The principle of the flexible spacer[10] (e.g.
-$(CH_2)_n$- or -$(CH_2)_nO$-) is vital in the field of LC polymers.
When such spacers are interposed between the mesogenic
groups of the main chain type or between the mesogenic
groups and the flexible backbone of the side chain type,
then LC phases are readily formed under appropriate
conditions for all classes of mesogenic group (rods, discs
or amphiphiles or mixtures of these in copolymers). In
other words, decoupling of the mesogenic groups from the
motions of the backbone, which tend to give a statistically
random conformation to the macromolecule, allows the

mesogenic groups to adopt the ordered arrangements (N, $S_C^*$, D, etc.) already defined for LMM LC materials.

In the absence of flexible spacer units in side chain LCP, the rigid or direct connections of the mesogenic groups to the backbone militate against LC phase formation, and the polymers are normally amorphous in character.

With no spacer units in a main chain system, the macromolecule is effectively a rigid rod. Such rigid rod polymers usually have high melting points, well above the temperatures at which rapid thermal decomposition occurs, and thermotropic LC phases are not observed. However, solutions of such polymers in rather specialised solvents such as sulphuric acid, polyphosphoric acid, etc. are anisotropic. Examples of such polymers are provided by poly-p-benzamide and poly-(p-phenyleneterephthalamide), and materials such as these have commanded great commercial interest because fibres drawn from the anisotropic solutions, followed by annealing, have very high tensile strength, e.g., Kevlar.

The molecular rods in such polymers vary greatly in length as the samples are polydisperse. A nematic ordering of such rods is not difficult to envisage, but a lamellar, smectic packing is hardly conceivable for rods with differing dimensions. The solutions are therefore nematic, and historically it is interesting to note that the chiral, rigid rod polypeptides studied by Robinson[11] in the 1950's, and then considered simply as lyotropic liquid crystals forming N* phases with solvents such as $CH_2Cl_2$, were some of the first examples of rigid rod polymeric LC materials. Solutions of RNA and tobacco mosaic virus are similarly anisotropic and liquid crystalline.

Polymers with Flexible Linkages

Such polymers may be depicted generally as:

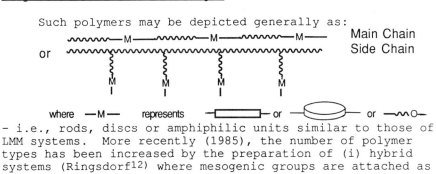

- i.e., rods, discs or amphiphilic units similar to those of LMM systems. More recently (1985), the number of polymer types has been increased by the preparation of (i) hybrid systems (Ringsdorf[12]) where mesogenic groups are attached as side chains to either the -M- groups or the flexible spacers of main chain polymers, and (ii) side chain polymers (Finkelmann[13]) in which rod-like mesogenic groups are attached laterally to the backbone via flexible spacers linked to the sides of the rods. In 1989, other hybrid systems[14] were reported in which rod-like mesogenic units are incorporated as side chains, some being terminally and some laterally linked by the spacer to the backbone. An example of such a statistical copolymer is (**I**).

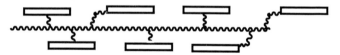

(**I**)

It must be stressed that for such a statistical lateral/ terminal side chain copolymer the intra- and inter-molecular parallel alignment of the side chains is probably achieved by adoption of a mesogen-jacketed structure such as

The side chain homopolymer (**I**, $x \approx 40$, $y = 0$) allows an important point to be made. Despite the long alkyl chains at each end of the rod, the polymer is purely nematic (Tg 4.8°C N 58°C Iso). Smectic phases have never been observed for such lateral homopolymers, and indeed, a lamellar packing of side chains is hard to conceive.

Side chain polymers with terminally attached mesogenic groups are however often smectic in character. For example, structure (**I**, $x = 0$, $y \approx 40$) represents a crystalline polymer that is purely smectic - K 107°C $S_A$ 184°C Iso). Introduction of as little as 15% of laterally attached side chain however eliminates smectic properties, and the side chain copolymer is purely nematic (e.g. **I**, $y + x \approx 40$; $y:x = 3:1$, Tg 5.8°C N 96.7°C Iso). Such hybrid copolymers allow phase behaviour and transition temperatures to be fine tuned with respect to other physical properties such as $\Delta\varepsilon$.

The structural diversity possible for such main chain and side chain polymers with flexible linkages is clearly very extensive. Mesogenic units may be selected from any of the three classes (rods, discs or amphiphiles) and incorporated as main chain and/or side chain functions. More than one structural type of rod, disc or amphiphile may be used or combinations of rods, discs and amphiphiles. Some of the units may be chosen to be less mesogenic, to cause deviations of the main chain from linearity and so decrease $T_g$ and other transition temperatures, or to be non-mesogenic side groups, e.g., alkyl groups to fine tune thermal and other properties. Side chain polysiloxanes prepared from poly(hydrogenmethyl-dimethylsiloxane) backbone provide an example of the latter type.

With such flexible spacers, both main chain and side chain LCP involving each of the three main types of mesogenic group have been made and shown to exhibit liquid crystal phases - calamitic, discotic or lyotropic. The ability to incorporate a proportion of units other than mesogenic units also permits functionalisation of LCP, e.g., by introducing chromophoric, electro-active etc. units.

## 3 MAIN CHAIN LCP WITH FLEXIBLE SPACERS

Formation of liquid crystal phases is favoured by homopolymers where the rods or discs are all the same and the spacers of equal length. The formation of N, N*, $N_D$, S and D phases is readily visualised; N phases are favoured by short spacers and D or S phases by longer spacers. Not surprisingly, statistically irregular polymers of this type (varying length of rods and/or spacers) favour N phases.

Some examples of flexibly linked main chain LCP are:

$$\left[-\bigcirc-CH=CH-\bigcirc-CO_2(CH_2)_6O_2C-\right]_n$$

$$\left[-\bigcirc-N=N-\bigcirc-O(CH_2CH_2O)_4-\right]_n$$

$$\left[-\bigcirc-\bigcirc-\bigcirc-CO_2(CH_2CH_2O)_{10}-\right]_n$$

## Identification of LC Phases

**Differential Scanning Calorimetry.** Heating and cooling cycles allow Tg and mesophase transition temperatures ($T_{N-Iso}$, $T_{SA-N}$, $T_{SA-Iso}$ etc.) to be identified (as reversible processes). Melt transitions (Tm) involving crystalline phases can also be evaluated.

**Optical Microscopy.** Liquid crystal phases may be identified by the optical textures of thin films viewed between crossed polarisers. Long annealing of the sample is often needed before characteristic textures[15] are adopted.

**X-ray Diffraction.** This technique is of great importance for confirming phase type through measurement of layer spacings and determination of lamellar order.

Flexibly linked main chain polymers are commonly made by polycondensation or trans-esterification reactions, and are as a consequence usually polydisperse. Like impure LMM systems or mixtures of LMM mesogens, they therefore show spread transitions. $T_{N-Iso}$ temperatures ranging over 10°C are not for example uncommon. The extent of the biphasic gap is all too rarely reported, but attention to this has been paid by some, notably Blumstein[16] working with carefully fractionated and characterised materials. It has also been demonstrated[17] that for such materials, regular alternations of mesophase transition temperatures occur as spacer length is regularly increased. Transition temperatures like Tm, Tg, $T_{N-Iso}$ etc are a regular function of degree of polymerisation ($\overline{DP}$) and cannot be assumed to be constant until $\overline{DP} > 30$ (equivalent to $\overline{M}_n \approx 14,000$).

## Applications and Rheology

These features depend of course upon the macroscopic alignment possible with LCP and have been reviewed by Cogswell[18]. Like rigid rod polymers, flexibly linked main chain polymers form high modulus, high tensile strength fibres when drawn from the nematic phase, e.g., polyethylene terephthate/p-hydroxybenzoic acid copolymer. They may also be used to obtain highly anisotropic mouldings, resins and extrudates; the anisotropy confers good stiffness and strength on the products, the low viscosity of the LCP enables products to be made from very complex mouldings, and the low coefficient of thermal expansion is another important factor. For these types of application, it can probably be said that variations in the thermal and other physical characteristics of the polymers with $\overline{DP}$ and $\overline{M}_w/\overline{M}_n$ (polydispersity) are not too critical.

## 4 SIDE CHAIN LCP WITH FLEXIBLE SPACERS

With suitable side chains or groups, these polymers are capable of forming enantiotropic liquid crystal phases either on heating above their Tg values or with amphiphilic side chains on addition of controlled amounts of appropriate solvent. With disc like pendent groups, $N_D$ and D phases may be produced thermally, and with calamitic groups, N, N* and a variety of polymorphic S phases may similarly be formed. In these thermotropic liquid crystal phases, the flexible backbone must be assumed to thread its way throughout the ordered arrangement of side chains, in the case of calamitic smectics, possibly lying to a considerable extent in the planes between layers.

Differential scanning calorimetry, polarising optical microscopy and X-ray diffraction are again the main techniques used to establish transition temperatures, phase types and phase structure.

Backbones

Siloxane and acrylate type backbones are commonly used, and much work has been carried out on such polymers. However, increasingly polyphosphazenes, polyalkene sulphones and malonate/diol derived polyesters carrying pendant mesogenic groups are being studied. The structures of these side chain LCP are shown below where M = mesogenic unit:

Polyacrylates and Polymethacrylates

$$\begin{array}{cc} -[CH_2-CH]_n- & -[CH_2-CMe]_n- \\ | & | \\ CO & CO \\ | & | \\ O & O \\ | & | \\ (CH_2)_m & (CH_2)_m \\ | & | \\ M & M \end{array}$$

(Tg often > 40°C)

Polysiloxanes (end-capped)  (Tg often ≤ Room Temperature)

$$Me_3Si-[O-SiMe]_n-OSiMe_3$$
$$\qquad\qquad\;\; |$$
$$\qquad\quad (CH_2)_m$$
$$\qquad\qquad\;\; |$$
$$\qquad\qquad M$$

Homopolymer

$$Me_3Si-[O-SiMe]_n------[O-SiMe_2]-OSiMe_3$$
$$\qquad\qquad\;\; |$$
$$\qquad\quad (CH_2)_m$$
$$\qquad\qquad\;\; |$$
$$\qquad\qquad M$$

Statistical Copolymer

Polyphosphazenes

$$\left[\begin{array}{c} R \\ | \\ P = N \\ | \\ (CH_2)_m \\ | \\ M \end{array}\right]_n$$

Polyalkene Sulphones

$$\left[\begin{array}{c} CH_2-CH-SO_2 \\ | \\ (CH_2)_m \\ | \\ M \end{array}\right]_n$$

Malonate/Diol Polyesters

$$\left[\begin{array}{c} O_2C-CH-CO_2-(CH_2)_x \\ | \\ (CH_2)_m \\ | \\ M \end{array}\right]_n$$

The diol moiety may also carry a mesogenic group either with or without a mesogenic group on the malonate moiety.

Although comparisons are valid only for materials of low $\overline{M}_w/\overline{M}_n$ and closely similar $\overline{DP}$ values (see later), the following data[19] permit methacrylate, acrylate, siloxane and siloxane copolymer backbones to be compared.

| Structure | Phase Transitions | Phase Range |
|---|---|---|
| $\left[CH_2-CMe(CO_2R)\right]$ | g 369°C N 394°C Iso | 25°C |
| $\left[CH_2-CH(CO_2R)\right]$ | g 320°C N 350°C Iso | 30°C |
| $\left[O-SiMe(CH_2R)\right]$ | g 288°C N 334°C Iso | 46°C |
| $\left[O-SiMe(CH_2R)\right]\left[SiMe_2\right]$ | g 276°C N 294°C Iso | 18°C |

where R = $-(CH_2)_2-O-\text{\textbenzene}-CO_2-\text{\textbenzene}-OMe$

Methacrylates have higher Tg and nematic thermal stabilities than acrylates. The very flexible siloxane backbone gives still lower Tg values and nematic thermal stabilities, and when only a proportion of the silicons in the backbone carry pendent mesogenic groups, still lower Tg (sub-ambient in the case above) $T_{N-Iso}$ values are observed.

The backbone therefore has an important influence upon the thermal properties of the LC system, and the mesomorphic properties of such polymers must be understood to be influenced not only by the mesogenic group, but also by the backbone and the length and nature of the spacer group. Although decoupled from the backbone, the mesogenic group is still influenced by the flexible spacer and the backbone.

### Diversity of Structure for SCLCP

It is clear from the foregoing discussion, the wide choice of mesogenic group(s) and the structures shown that there is a very large diversity of structure possible for flexibly linked side chain liquid crystal polymers, and much interest in these systems, both academic and commercial, has been generated in recent years. There are three reasons.

(1) SCLCP are novel materials and studies of their physical properties can enhance our understanding of the fundamental science of both polymers and liquid crystals.
(2) It is a worthy objective to seek to establish a clearer understanding of the relations between SCLCP structure and thermodynamic properties.
(3) Following upon the successes of LMM LC materials in the area of technological applications, LCP have been viewed with especial interest for their potential for applications, particularly in the fields of optics and optoelectronics.

This last reason has in fact formed a very strong motivating and driving force for the study of SCLCP and much research on these systems is commercially driven. Some attention will now be paid to possible applications for such macromolecular materials. As will be seen, these applications are rather sophisticated, often depending critically on quite precise physical parameters of the polymers. The consequences of this will then be considered for the chemist who must make reproducible products, first on a laboratory scale and later on a major scale if market needs for eventual commercial devices are developed.

### Possible Areas of Application for SCLCP

**Fixed Wavelength Light Filters and Reflectors** A thin film of a cholesteric (N*) liquid crystal of low molar mass mounted between flat glass substrates and aligned such that the long axis of the helical structure is uniformly orthogonal to the surfaces often exhibits angle dependent, selective reflection of coloured light. The wavelength of light reflected lies in a fairly narrow wavelength band and is dependent on the length of the pitch and therefore on temperature. At a given angle of view, the colour of the film may therefore change as the temperature falls from the ultraviolet through the entire visible spectrum into the infrared. Hence the thermochromic applications of LMM LC[20].

A similarly aligned sample of a chiral SCLCP behaves in the same manner. Here however, we have the glassy state to consider. Once a liquid crystal phase converts into the glass, the orientational order of the phase is frozen into the glass. With an N* phase, the cholesteric helix therefore becomes locked into the glassy state. If therefore we use temperature to tune the pitch and the wavelength of selective reflection to a particular value ($\lambda_T$) and then rapidly quench the film, the film of glass will permanently reflect that wavelength of light and exhibit a particular colour as long as Tg is not exceeded. Cast films may therefore behave as fixed wavelength filters or reflectors.

Finkelmann[21] was the first to realise these possibilities and to demonstrate that the pitch and selective reflection properties of SCLCP can also be controlled structurally by appropriate molecular engineering of the side chain homopolymer or copolymer. Using a siloxane side chain copolymer incorporating two different side chains, one chiral (**II**) where chol = cholesteryl and one not (**III**) in different

—$(CH_2)_3CO_2$ Chol      (**II**)

—$(CH_2)_mO$ —⟨⟩—$CO_2$—⟨⟩—OMe   (**III**)

relative molar amounts, he showed that the pitch in a polymer is tighter than would be expected from a similar mixture of the LMM equivalents of the side chain precursors. As anticipated, as the molar proportion of (**II**) relative to (**III**) in the copolymer was increased, the pitch tightened as it also did for a constant ratio of (**II**) to (**III**) as the spacer length m in (**III**) was reduced. This affords wide possibilities for tuning the optical properties of cast films.

A development[22] which extends these possibilities involves mesogenically substituted derivatives of cyclic (hydrogenmethyl)siloxanes, e.g.,

$$\left[\begin{matrix} \text{Me---Si---(CH}_2)_3\text{O Chol} \\ \quad\;\; | \\ \quad\;\; \text{O} \end{matrix}\right]_x \quad \text{(where } x + y = 5)$$

$$\left[\begin{matrix} \text{Me---Si---(CH}_2)_3\text{O---}\langle\!\bigcirc\!\rangle\text{---CO}_2\text{---}\langle\!\bigcirc\!\rangle\text{---}\langle\!\bigcirc\!\rangle \\ \quad\;\; | \\ \quad\;\; \text{O} \end{matrix}\right]_y$$

Such materials form good pliable surface coatings on paper which reflect iridescent cholesteric colours, varying with viewing angle, in the glassy state (blue to red with decreasing angle). The colour/angle dependence can be reversed by addition of suitable pigments.

## Electrically/Thermally Addressed Smectic A storage Device[23]

An unaligned film of a terminally attached side chain polymer (Tg-$S_A$-Iso) with at least a proportion of the side chains substituted such that they are of positive dielectric anisotropy will be light scattering (layer curvature in the focal-conic $S_A$ state). If dichroic dyes have been incorporated in the polymer as additional pendant groups or simply as a LMM dopant which aligns with the side chains, the scattering film will be coloured (or black) according to the dye or dyes used. If the glass plates are coated with a thin transparent film of tin oxide/indium oxide, a voltage may be applied across the film. When this was done at 75-85°C for the particular polymer studied – this temperature range representing the biphasic gap for the $T_{S_A\text{-}Iso}$ transition – the side chains became aligned perpendicular to the plates. Information could therefore be written into the cell electrically, as the aligned regions would be clear against a scattering background or colourless against the coloured dyed background.

| Unaligned Polymer | 200 V; 75-85°C → | Aligned written areas |
|---|---|---|
| Scattering or coloured | ← Heat > 110°C Cool | Clear/colourless areas |

Erasure was achieved by heating the whole cell above $T_{S_A\text{-}Iso}$ and cooling with field off. The written information could be stored very durably in the smectic state if Tg was below ambient, or obviously in the glassy state with higher Tg materials. With frequency optimization, switching at 200 ms was claimed, the contrast was quoted as greater than that for the LMM analogous device, achievement of grey scale was possible by control of pulse width, voltage and frequency, and the device also functioned with a negative $\Delta\varepsilon$ polymer.

In this device, the writing process is conducted in the biphasic gap where viscosity effects are minimised, so overcoming the problem that viscosity effects will not normally allow SCLCP to compete with their LMM analogues for use in fast switching electro-optical displays operating at ambient temperatures. Note that success for such a device requires a polymer with reproducible transition temperatures.

Laser or Thermally Addressed Storage Devices[24]

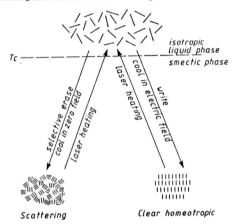

As shown above, this device works again on the principle of starting in the unwritten state with a scattering or coloured (dyed) film of the smectic phase of the LC polymer in an unaligned state. Using a computer-controlled laser to heat the polymer locally above its isotropisation temperature and cooling as the beam tracks on, information may be written in through the action of an applied field which aligns the written areas homeotropically so that they appear clear or colourless against the scattering or coloured background. Partial or selective erasure is achieved by laser writing the areas with no field applied and total erasure by heating the cell above the clearing temperature of the mesophase and cooling with the field off.

Starting with the aligned, homeotropic smectic phase (by cooling the entire isotropic film in an electric field) positive contrast writing can be achieved by laser heating and cooling with the field off - and erasure by rewriting with the field on.

The possibilities for such devices in optical data storage have been thoroughly reviewed by McArdle[24], including write once and erasable polymeric recording media. Laser heating in such devices could of course be replaced by a matrix of heating elements, giving a dot matrix data presentation. In all such devices, the written data could be stored either in the smectic phase if Tg is below ambient or more durably in the glass phase if higher Tg polymers are used.

A striking example[25] of the resolution achievable using laser writing on a smectic polysiloxane was reported in 1987 and illustrated by a portion of an ordnance survey map. The positive $\Delta\varepsilon$ polymer used in this case was

$$Me_3Si-O{\left[-SiMe-O\right]}_x{-----\left[-SiMe_2O-\right]}_y-SiMe_3$$
$$(CH_2)_6$$
$$O-\langle\_\rangle-CO_2-\langle\_\rangle-CN \quad Tg\ -12°C\ S_A\ 100°C\ Iso$$

$x:y = 1; x + y \approx 35$

and the written information was stored in the $S_A$ phase.

Developments in this area continue apace, activity centring around flexible, SCLCP films incorporating infrared dyes and capable of giving good contrast and grey scale. Results obtained under an EEC/BRITE project (GEC, Akzo International and the Universities of Hull and Leeds) are to be reported at Eurodisplay 90[26].

## Ferroelectric/Pyroelectric SCLCP

The unwound tilted $S_C$ phase of a chiral LMM LC material in which a strong lateral dipole in the rod shaped molecule is located close to or at the chiral centre exhibits a macroscopic spontaneous polarisation ($P_S$) and is ferroelectric and thus pyroelectric. Much effort has been given to studies of electro-optic display devices based on such materials and capable of µs switching speeds. Such devices therefore switch much faster than LMM LC nematic displays (ms) and much interest was created by the report[27] that a chiral polymer with a ferroelectric $S_C^*$ phase had a very large $P_S$ of 1,000-2,000 nC cm$^{-2}$ and gave sub-millisecond switching, albeit at about 60°C. The polymer was a polyacrylate with the side chain

$$-(CH_2)_{10}O-\bigcirc-CO_2-\bigcirc-\bigcirc-O_2C\overset{*}{C}H\overset{CH_3}{\underset{Cl}{|}}\overset{*}{C}HC_2H_5$$

Ferroelectric SCLCP may yet therefore give workable fast switching displays. More work needs to be done in this area with this objective in mind, and also on that of using such polymer films in pyroelectric sensors or detectors.

## Side Chain LC Polymers as Optically Non-Linear Media

**Electric Field Poled Systems.** Since space is limited and the achievement of second order nonlinear effects by electric field poling of initially isotropic and liquid crystalline systems has been extensively reviewed by Mohlmann and van der Vorst[28], the reader is referred to this source and here we will consider the possibilities of achieving polar order through Langmuir-Blodgett films.

**Langmuir-Blodgett Films of Amphiphilic SCLCP.** Two examples of such polymers are given, but it must be emphasised that research is still in its early stages. The first type has been the vinyl-maleic anhydride copolymers (**IV**) of Hodge et al.[29] and the second, backbone copolymeric siloxanes[30] (**V**).

(**IV**) structure with R, HO$_2$C, CO$_2$Z groups, subscript n

(**V**) $Me_3Si\!-\!\!\left[O\!-\!\underset{R}{SiMe}\right]_x\!-\!-\!-\!-\!\left[O\!-\!SiMe_2\right]_y\!-\!OSiMe_3$    (x:y ≈ 1; x + y ≈ 26)

In (**V**), the mesogenic side chain was

R = -(CH$_2$)$_6$O-⟨◯⟩-N=N-⟨◯⟩-CH$_2$CH$_2$OH    (**VI**)

with a net dipole towards the hydrophilic group; stable monolayers and LB stacks of over a 100 layers were readily formed. The deposition was however Y type and although the monolayer showed evidence of second harmonic generation, dipole cancellation occurred in the multilayer stack. With

R = -(CH$_2$)$_6$O$_2$C-⟨◯⟩-N=N-⟨◯⟩-OCH$_2$CH$_2$OH  (**VII**)

the permanent dipole is directed away from the hydrophilic group. Again stable monolayers and extensive LB stacks can be achieved, but deposition is again Y type. Interest therefore resides in forming alternating LB stacks using the two SCLCP (**V**) with side chains (**VI**) and (**VII**) respectively. With Y-type deposition, dipole reinforcement and a macroscopic polarisation should result. At present there is evidence that this is possible and that such alternating stacks of appropriate polymers can give enhanced second harmonic generation.

Much work remains to be done in this area and little is known about the stability of such alternating stacks with particular reference to diffusive effects between layers.

Photochromic Polymers

Cabrera, Krongauz and Ringsdorf[31] first envisaged functionalisation of SCLCP by incorporating photochromic spiropyran moieties (**VIII**) as a proportion of the pendant side chains. They used a polysiloxane with side chains:

-(CH$_2$)$_6$O—⟨⟩—CO$_2$—⟨⟩—OMe

and -(CH$_2$)$_{10}$CONH—⟨Me,Me⟩—N⟨O⟩—⟨⟩—NO$_2$    (**VIII**)
                                    Me

The polymer had Tg ≈ 10°. A photochromic effect arises from the classical spiropyran-merocyanine interconversion. A cast film is strongly birefringent (pale pink at room temperature). Light of λ > 500 nm changes the colour to pale yellow (at room temperature) and light of λ = 365 nm changes this at room temperature to deep red and at -20°C to deep blue.

These observations point the way to the use of SCLCP in imaging technologies. Spiropyrans do however cause difficulties because of colour bleaching, and other photochromic systems and photochromic processes[32] are actively under consideration. Recent review references are quoted[33].

Elastomers for Integrated Optics and Membranes

The cross linking of SCLCP is an obvious possibility first realised by Finkelmann. A bifunctional cross linking agent is all that is required. Stretching of such an elastomer will give macroscopic alignment of the side chains as achieved for LMM LC materials using applied fields. An interesting possibility was reported[34] for a film of such an elastomer with the direction of orientation of the long axes of the terminally appended mesogenic side chains orthogonal to the plane of the film. An embossed die pressed on top of such a film will cause local distortions of the director above Tg, and below Tg these will be frozen in as light conducting pathways of as little as μm dimensions. This has clear implications for elements for integrated optics.

Finkelmann has also drawn attention to the selectivity of elastomer membranes with respect to permeation by molecules of different shapes. Spherical molecules of neopentane diffuse very much more slowly than elongated molecules of the isomeric n-pentane which is more compatible with the parallel aligned structure of the LCP. Possibilities therefore exist for membranes for molecular separations.

## 5 SYNTHETIC PROBLEMS RELATING TO SCLCP FOR APPLICATIONS

The applications of SCLCP discussed in section 4 are obviously quite sophisticated, and in many cases the end device can only function consistently if polymers with reproducible properties are made from batch to batch. In addition to transition temperatures, consistency of viscosity, dielectric properties, elastic constants etc. is clearly important. Yet we know that Tg values, isotropisation temperatures for mesophases, and the extent of biphasic gaps are strongly dependent upon $\overline{DP}$ and $\overline{M}_w/\overline{M}_n$, which are often variables for different polymer preparations

### Polysiloxanes

It has been reported that the transition temperatures for oligomeric siloxanes with mesogenic side chains rise steeply at first, but that after a degree of polymerisation of about 10, transition temperatures are pretty well constant. SCLC polysiloxanes are made by a Pt catalysed hydrosilylation reaction between a mesogenically substituted terminal alkene and a preformed backbone. These backbones are commercially available with varying $\overline{DP}$ values and polydispersities. For the homopolymer backbone, a common $\overline{DP}$ value would be about 40, and it might be thought that the resulting polymers would have transition temperatures lying on the flat portion of the T v $\overline{DP}$ plot and that results would be relatively unaffected if DP subsequently fell somewhat below or greatly exceeded 40. However, results obtained[35] using narrow molar mass fractions (preparative GPC) of poly(hydrogenmethylsiloxane) backbone of low $\overline{M}_w/\overline{M}_n$ (1.06-1.15) mesogenically substituted with

$$-(CH_2)_6O-\!\!\left\langle\!\!\bigcirc\!\!\right\rangle\!\!-CO_2-\!\!\left\langle\!\!\bigcirc\!\!\right\rangle\!\!-CN$$

show that Tg and $T_{SA-Iso}$ values are still rising significantly up to $\overline{DP}$ 100 and are only likely to level at $\overline{DP} > 150$. From $\overline{DP} = 40$ to $\overline{DP} = 107$, $T_{SA-Iso}$ rose by 17°C and Tg by 12°C.

Similar experiments with the lateral side chain

$$C_8H_{17}O-\underset{}{\underset{O(CH_2)_n-}{\bigcirc}}-CO_2-\bigcirc-\bigcirc-CN \qquad (IX)$$

show[36] that Tg and $T_{N-Iso}$ values can again be considered to be constant within experimental error only when $\overline{DP} > 150$.

Transition temperatures also vary with change in $M_w/M_n$, and this means that reproducible preparations of SCLC polysiloxanes can only be achieved using prepolymer backbone that has been checked and monitored for consistency of $\overline{DP}$ and $\overline{M}_w/\overline{M}_n$. This is especially true if $\overline{DP} < 150$ which is usually the case for commercially available prepolymer. However, possibilities may exist for the use of pure poly-(cyclosiloxanes) which under controlled acidic or basic

$$\left[\begin{array}{c} R \quad R_1 \\ \backslash / \\ Si-O \\ \end{array}\right]_n \text{ cyclic} \qquad \begin{array}{l} R = Me; R_1 = H \\ R = Me; R_1 = \text{mesogenic group} \end{array}$$

conditions give ring opening that could be followed by anionic polymerisation with BuLi as initiator to yield narrow polydispersity and controlled high DP open chain polymer.

A further factor concerns the hydrosilylation process itself[37]. With calamitic pendant groups attached terminally to the backbone, addition of the SiH to the terminal alkene usually proceeds normally to give $Si-CH_2-CH_2(CH_2)_n-$, at least when $n + 2 \geq 4$. However, with laterally attached side chains such as (IX), high percentages (up to 30%) of anomalous addition can occur[38] even with long spacers when $H_2PtCl_6$ is used as catalyst. This is avoided using divinyltetramethyldisiloxane Pt complex as catalyst.

Acrylates/Methacrylates

Here again, high DP values are required (> 100). High $\overline{DP}$ polymers are readily enough prepared[10] by traditional radical catalysed reactions in solution when methacrylates are involved, but acrylates normally give low $\overline{DP}$ polymers because of chain transfer reactions. Even with methacrylates, the problem of polydispersity remains.

An answer to these difficulties might seem to be provided by the procedure of Webster et al. from du Pont, first announced[39] in 1983. This involves what is known as group transfer polymerisation of methacrylate monomers and other activated α-alkenes. It is initiated by a ketene ketal $Me_2C=C(OR)OSiMe_3$ and catalysed by fluorides such as $nBu_4NF$, $KHF_2$ and tris(dimethylamino)sulphonium bifluoride.

Reactions are conducted at room temperature and the polymers have a narrow polydispersity ($\bar{M}_w/\bar{M}_n < 1.3$). Molecular weights are controlled by the ratio of monomer to initiator and mostly were in the range $\bar{M}_n = 1,000-2,000$. The polymers were effectively living polymers making possible the production of block copolymers by adding a fresh monomer at the end of the polymerisation. Acrylate esters were reported to give problems. However, in our hands, these methods with mesogenically substituted acrylates and methacrylates have been unsuccessful, but other catalysts await trial.

Developments reported by Reetz[40] may be helpful. This involves methods for the polymerisation of *acrylates* (and methacrylates) by a room temperature group transfer polymerisation process using carbon nucleophiles – tetrabutylammonium malonates – as initiators. These are stable and the $\overline{DP}$ is controlled again by the initiator:monomer ratio. Controlled values of $\overline{DP}$ up to 20,000 were obtained with $\bar{M}_w/\bar{M}_n$ values $\leq 1.3$. Again the polymers are living polymers. It remains to be seen whether the procedures work with mesogenically substituted acrylates and methacrylates.

Methods involving radical and pressure catalysed polymerisation of acrylates/methacrylates using different solvents may produce suitable SCLCP; also radical procedures promoted by sonication and pressure should be investigated. Other possibilities such as starting with monodisperse polyacrylic acid (commercially available) and its derivatives, e.g., polyacryloyl chloride or sodium polyacrylate, and appending the mesogenic moieties to the preformed backbone seem less attractive.

Malonate/Diol Polyesters

These are prepared by transesterification and DP values are low (10-20). Polymers probably therefore relate to a steep part of any physical parameter $v$ $\overline{DP}$ plot. A similar problem exists for polyvinyl ethers reported by Percec[41].

6 FUTURE DEVELOPMENTS

From what has gone before, it is clear that potential applications for liquid crystal polymers exist. Those for main chain polymers are almost certainly already well under commercial scrutiny and assessment. The question marks would seem to relate to the more specialised applications for side chain liquid crystal polymers and the problem is that if strictly reproducible batches of polymers must be made to ensure consistency of the devices, do the chemical procedures exist to do this reliably and reproducibly? If they do not, then the answer can only be provided by some fundamental studies aimed at methods of making defined DP, low polydispersity mesogenated SCLCP available at a viable cost. It is too early yet to talk of market requirements for these materials, but it is certainly true that, as in so many of today's "high tech" developments, the key to success can only be produced by chemists.

REFERENCES
1. A. Blumstein (ed.), 'Polymeric Liquid Crystals', Plenum, New York, 1985.
2. L. Chapoy (ed.), 'Recent Advances in Liquid Crystalline Polymers', Elsevier, London, 1985
3. A Ciferi, W. Krigbaum and R. Meyer (eds.), 'Polymer Liquid Crystals', Academic Press, New York, 1982.
4. C. McArdle (ed.), 'Side Chain Liquid Crystal Polymers', Blackie, Glasgow, 1989.
5. M. Cox, 'Liquid Crystal Polymers', RAPRA Report No. 4, Pergamon, Oxford, 1987.
6. V. Shibaev and N Platé, Adv. Polym. Sci., 1984, 60/61, 173.
7. G.W. Gray (ed.), 'Thermotropic Liquid Crystals', CRAC series, Vol. 22, Wiley, Chichester, 1987.
8. G.W. Gray and P.A. Winsor (eds.), 'Liquid Crystals and Plastic Crystals', Vols. 1 and 2, Ellis Horwood, Chichester, 1974.
9. S. Chandrasekhar, 'Liquid Crystals', University Press, Cambridge, 1980.
10. H. Finkelmann in 'Thermotropic Liquid Crystals' (G.W. Gray, ed.), CRAC Series, Vol. 22, Chap. 6, Wiley, Chichester, 1987.
11. C. Robinson, J.C. Ward and R.B. Beevers, Discuss. Faraday Soc., 1958, 25, 29.
12. B. Reck and H. Ringsdorf, Makromol. Chem. Rapid Commun., 1985, 6, 291.
13. V. Hessel and H. Finkelmann, Polymer Bull., 1985, 14, 3751.
14. M.S.K. Lee, G.W. Gray, D. Lacey and K.J. Toyne, Makromol. Chem. Rapid Commun., 1989, 7, 71.
15. G.W. Gray and J.W. Goodby, 'Smectic Liquid Crystals', Leonard Hill, Glasgow, 1984.
16. A. Blumstein, Polym. J., 1985, 17, 277.
17. R.W. Lenz, Polym. J., 1985, 17, 105.
18. F.N. Cogswell, 'Recent Advances in Liquid Crystal Polymers', (L. Chapoy, ed.), Chap. 10, Elsevier, London, 1985.

19. H. Finkelmann and G. Rehage, Adv. Polym. Sci., 1984, 60/61, 101.
20. D.G. McDonnell in 'Thermotropic Liquid Crystals' (G.W. Gray, ed.), CRAC Series, Vol. 22, Chap. 5, Wiley, Chichester, 1987.
21. H. Finkelmann and H.J. Kock, Disp. Technol, 1985, 1, 81.
22. H.-J. Eberle, A. Miller and F.-H. Kreuzer, Proc. 12th Int. Liq. Cryst. Conf., Freiburg, FRG, 1988.
23. H.J. Coles and R. Simon in 'Polymer Liquid Crystals' (A. Blumstein, ed.) Plenum, New York, 1985.
24. C. McArdle in 'Side Chain Liquid Crystal Polymers' (C McArdle, ed.), Chap. 13, Blackie, Glasgow, 1989.
25. C. McArdle, M. Clark, C. Haws, M. Wiltshire, A. Parker, G.W. Gray, D. Lacey and K.J. Toyne, Liq. Cryst., 1987, 2, 573.
26. C. Bowry, et al. Proc. Eurodisplay 90, to be published.
27. H. Gleeson, H.J. Coles and G. Sherowsky, Proc. British Liq. Cryst. Soc., Sheffield, 1989.
28. G.R. Mohlmann and C.P.J.M. van der Vorst in 'Side Chain Liquid Crystal Polymers' (C. McArdle, ed.), Chap. 12, Blackie, Glasgow, 1989.
29. C.S. Winter, R.H. Tredgold, A.J. Vickers, E. Khoshdel and P. Hodge, Thin Solid Films, 1985, 134, 49.
30. N. Carr. M. Goodwin, A. McRoberts, G.W. Gray, R. Marsden and R.M. Scrowston, Makromol. Chem. Rapid Commun., 1987, 8, 487.
31. I. Cabrera, V. Krongauz and H. Ringsdorf, Mol. Cryst. Liq. Cryst., 1988, 155, 221.
32. E. Lemaitre, X. Coqueret, R. Mercier, A. Lablache-Combier and C. Loucheux, J. Appl. Polym. Sci., 1987, 33, 2189.
33. H. Durr, Angew. Chem. Int. Ed. Engl., 1989, 28, 413; H.-W. Schmidt, Angew Chem. Int. Ed. Engl. Adv. Mater., 1989, 28, 940.
34. H. Finkelmann, Angew Chem. Int. Ed. Engl., 1988, 27, 987.
35. G.W. Gray, W.D. Hawthorne, J.S. Hill, D. Lacey, M.S.K. Lee, G. Nestor and M.S. White, Polymer, 1989, 30, 964.
36. M.S.K. Lee and G.W. Gray, unpublished results.
37. G.W. Gray in 'Side Chain Liquid Crystal Polymers' (C. McArdle, ed.), Chap. 4, Blackie, Glasgow, 1989.
38. G.W. Gray, J.S. Hill and D. Lacey, Mol. Cryst. Liq. Cryst. Lett., 1990, 7, 47.
39. O.W. Webster, W.R. Hertler, D.Y. Sugah, W.B. Farnham, T.V. Rajan-Babu, J. Amer. Chem. Soc., 1983, 105, 5706.
40. M.T. Reetz, Angew. Chem. Int. Ed. Engl., 1988, 27, 994.
41. V. Percec, Proc. British Liq. Cryst. Soc., Bristol, 1990.

# Polymers for Non-linear and Electro-optic Applications

R. N. DeMartino

HOECHST CELANESE RESEARCH DIVISION, ROBERT L. MITCHELL TECHNICAL CENTER, 86 MORRIS AVENUE, SUMMIT, NEW JERSEY 07901, USA

## 1 INTRODUCTION

Organic and polymeric materials have emerged in recent years as promising candidates for advanced device and systems applications.[1,2] This interest has arisen from the promise of extraordinary optical, structural, and mechanical properties of certain organic materials, and from the fundamental success of molecular design performed to create new materials.

Non-Linear Optics

Non-linear optics concerns the interaction of laser light with matter. This interaction causes a change in the material properties (refractive index, transparency) which can alter laser properties such as frequency and amplitude/phase. These changes can be exploited for various applications. These effects are very important because they now allow for the control of light by light and this may lead to optical devices which will be faster than electronic devices. Non-linear optical (NLO) devices can therefore have a major impact on the performance of laser communication, processing, and computing systems. This impact is twofold: NLO devices can improve the performance of existing components and systems, and lead to the design of new systems based on their unique properties. Table 1 highlights the applications at both device and systems levels.

Table 1  Non-linear Optical Applications

| Device Level | System Level |
|---|---|
| High Speed Modulators | Telecommunications |
| Optical Switches | Optical Computing |
| Waveguide Couplers | Signal Processing |
| Tunable Lenses | Telecommunications |
| Crossbar Switches | Laser Communications |

Basic Concepts of Organic NLO

The equation below describes the polarization of an organic molecule, as a function of the applied electric field E, and indicates the linear ($\alpha$) and non-linear ($\beta$, $\gamma$) terms:

$$\mu = \mu_o + \alpha E + \beta E^2 + \gamma E^3 + \ldots \qquad (1)$$

This equation depicts the molecular response. As chemists and material scientists, we are interested in transforming this activity into a bulk system so as to produce marketable devices. The equation describing this transformation is:

$$\chi^{(2)} = Nf <\beta> \qquad (2)$$

Where: $\chi^{(2)}$ is the bulk second order non-linear optical property; N, the number of molecules; and f, the local field factors. One can readily see that raising $\beta$ will result in a higher bulk activity. The operative mechanism for organic NLO materials is $\pi$ cloud excitation and subsequent electron motion. Increasing this probability will increase the $\beta$ of the molecule. $\beta$ depends on the difference in the ground and excited state dipole moments. By building an electronic asymmetry into the molecule (through the use of donor and acceptor groups), one can raise this difference and consequently increase $\beta$. Another important factor affecting $\beta$ is the transition dipole moment or the number of electrons traveling back and forth along the molecule. Thus, an efficient conduit between the donor and acceptor groups will also enhance $\beta$. In summary, high $\beta$

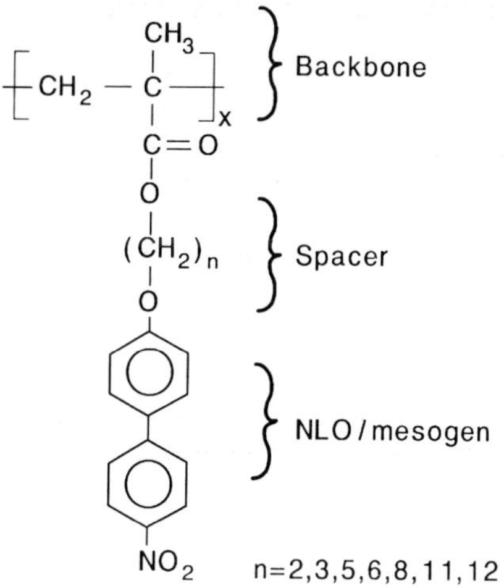

$$\left[ CH_2 - \underset{\underset{O}{\overset{|}{C}=O}}{\overset{CH_3}{\underset{|}{C}}} \right]_x \Big\} \text{Backbone}$$

$$\underset{O}{\overset{|}{(CH_2)_n}} \Big\} \text{Spacer}$$

[diphenyl ring system]
$\Big\}$ NLO/mesogen

$NO_2 \quad n=2,3,5,6,8,11,12$

**Figure 1**   First Generation NLO Polymers

requires an asymmetric potential with a large charge delocalization.

Polymer Advantages

However, β is a tensor quantity and, therefore highly symmetry dependent. Thus, to achieve bulk second order properties, it is also necessary to align the dipoles such that they all point in the same direction. Small molecule organics tend to crystallize with dipoles opposing each other to cancel out the charges. When this happens, bulk second order properties are lost. To circumvent this, active molecules can be incorporated into a polymeric structure, such as depicted in Figure 1. One can then heat the polymer above $T_g$, apply an electric field to align the dipoles, and subsequently cool to room temperature (below $T_g$) with the field on to freeze in the non-centrosymmetric structure.

For device applications, organic polymer systems offer a number of unique optical and structural advantages over other materials. They can achieve high optical quality and transmission over many wavelengths. Polymers are mechanically strong and stable materials capable of being fabricated into many shapes, sizes, and forms, including layered composite structures. A particularly distinct advantage lies in the ability, through appropriate chemistry, of engineering the molecular properties to achieve the essential macroscopic properties. Polymer chemistry can also tailor the critical secondary properties (glass transition temperature - $T_g$; phase type and clearing temperatures, absorption/transmission windows, etc.) that are extremely important for fabrication purposes to produce a viable product. When coupled with low refractive indices and dc dielectric constants, the cumulative properties of these materials show exceptional promise towards improving the performance of existing electro-optic and nonlinear optical devices.

To date, a significant materials synthesis effort has produced compounds that are approaching the levels of performance now achievable by current inorganics.[3] Many problems remain to be solved, including the realization of high optical quality, low loss, and acceptable long term thermal, oxidative, and photo-oxidative stability.

2 DEVELOPMENT OF SECOND ORDER POLYMERS

The general approach to the problem is to: 1) develop a molecular level understanding of what types of molecules possess high NLO activity; 2) incorporate the appropriate functionality to produce NLO active polymers; and 3) fabricate these materials for device evaluations. The primary goals one needs to assess, when considering a candidate polymer include: 1) NLO activity of the moiety; 2) optical transparency at the critical wavelengths; 3) long term stability of the poled state at specified operating temperature ranges; and 4) processability. All criteria must be satisfied before a new material can be utilized.

It was quickly realized that the activity of the oxy/nitro biphenyl unit (Figure 1) was not high

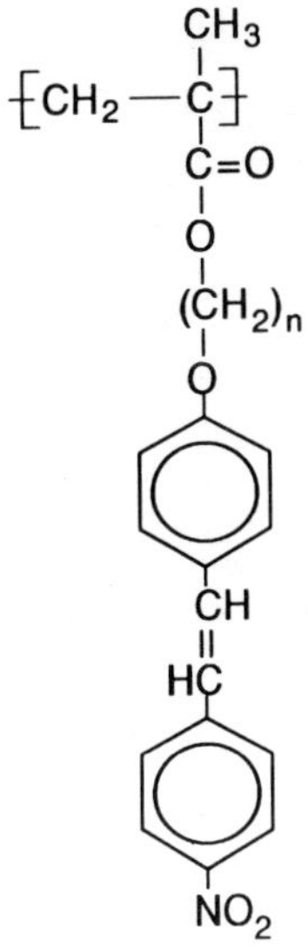

$$\text{\texteurohyphen}[CH_2-\underset{\underset{\underset{\underset{\underset{\underset{\underset{\underset{NO_2}{\bigcirc}}{|}}{HC}}{\overset{\|}{CH}}}{\bigcirc}}{\underset{|}{O}}}{\underset{|}{(CH_2)_n}}}{\overset{\overset{\overset{CH_3}{|}}{C}}{\underset{|}{\underset{|}{C=O}}}}]\text{\textendash}$$

Figure 2    ONS Based NLO Polymers

enough to warrant further investigation. Moreover, the presence of a liquid crystalline state might create additional scattering for wave guide applications. The next generation polymers were designed with these factors in mind.

Figure 2 shows the structure of the oxy/nitro stilbene (ONS) based polymers. Spacer lengths of 3, 6, and 11 were initially chosen. The first objective was to increase the $T_g$ and this could be realized through the use of comonomers. Figure 3 depicts the effect of increasing amounts of a comonomer, methyl methacrylate (MMA), on the $T_g$ of the resulting polymer. As expected, the $T_g$ increases with increasing MMA content. However, in the case of the 11 spacer polymer, a liquid crystalline phase still existed. This was avoided by utilizing shorter spacers in conjunction with a comonomer, as shown in Figure 4. Increasing amounts of MMA, for 11 spacer polymers, resulted in higher glass temperatures and less ordered LC states. Separating the mesogenic groups, with 50 mole % MMA, resulted in a nematic phase. Although a less ordered phase did develop,

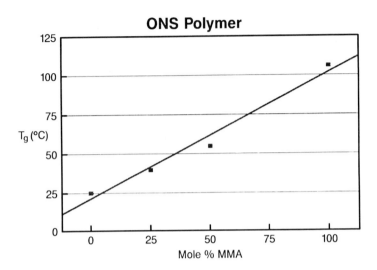

Figure 3    Comonomer Effect on Polymer $T_g$

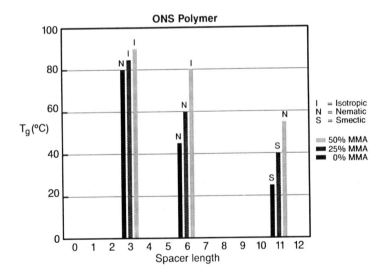

Figure 4    Comonomer Effect on $T_g$ and LC Phase

the side chain was long enough to partially overcome this separation, and an isotropic phase did not occur. As the spacers became progressively shorter, it became easier to form isotropic polymers with increasing amounts of MMA. Thus, 50% MMA produced isotropic polymers with a six spacer unit, while only 25% MMA was needed for the three spacer copolymer.[4]

All subsequent side chain copolymers, with even higher active units, were prepared in this manner, resulting in isotropic materials with high glass temperatures for long term stability. The present polymer, HCC-1232, incorporates an NLO unit with a measured $\beta$ of 71 x $10^{-30}$ esu and a $T_g$ in excess of 135°C. This results in stabilities approaching 5 years at operating temperatures of 70°C.

## 3 CHARACTERIZATION

There are both primary and secondary properties of organic materials that must be carefully understood and measured. Primary material properties include basic material parameters, such as magnitudes of the real and imaginary parts of the linear and nonlinear susceptibility over a broad frequency range, from dc to optical frequencies. Secondary properties, such as the optical scattering losses, optical quality, and thermal properties must also be measured for optimization of material synthesis procedures. It is not necessary to carry out all measurements for each material system. A preliminary measurement of the second order nonlinearity of a poled polymer film might show that the poling procedure requires modification to achieve greater orientation in the film, and thus the material synthesis needs to be further refined. Promising results for the poled film would lead to measurements of the frequency-dependent dielectric constant, absorption coefficient, and overall frequency response.

These measurements are essential for two reasons. First, ongoing synthesis work requires knowledge of the primary material parameters for optimization of various parameters and continuation of synthesis work, and requires knowledge of the secondary properties to point to changes in fabrication techniques to improve optical clarity, quality, and minimize absorption and scattering losses. Second, device concepts and performance parameters require accurate values of material parameters for meaningful evaluation of performance of a particular device with a particular material. The identification of a promising material system for a given device requires that the device performance be evaluated with that material's primary properties.

Before a polymeric material can be evaluated, it needs to be poled in an electric field to develop the bulk second order activity. This can be accomplished by first spin coating the polymer onto a suitable substrate. After the appropriate drying steps, the film is electroded and wired. The film is then heated to the glass temperature and an electric field applied. After a certain time period, the film is slowly cooled, with the field on, to room temperature. This process results in a permanently poled system.

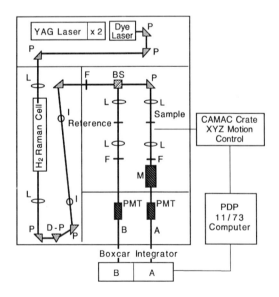

Figure 5   Second Harmonic Generation Testbed

## Second Harmonic Generation

The apparatus for the second harmonic generation studies is shown in Figure 5. The dye laser is pumped by a frequency doubled Yag laser to give wavelengths in the range of 500-800 nm. This is used to pump an $H_2$ filled Raman cell to extend the wavelength range by the stimulated Raman effect. After the appropriate wavelength is selected, it is split into two for the sample and reference arms. The poled polymer film sample is placed in the device at an angle of 0° relative to the laser beam. At this point, there is no second harmonic signal observed due to the fact that the aligned molecules are 90° relative to the electric field component of the laser beam. As the sample is rotated, the molecules start to interact with the E field and second harmonic signals are observed. The measurements are performed relative to a standard material such as quartz, BK7 glass, or $CS_2$. The energy in each pulse is integrated by a boxcar integrator, and each pulse from the sample arm is divided by the reference arm to compensate for laser pulse fluctuations. The translation or rotation of the samples are controlled by a computer which also averages and stores the data. A more complete description of the test can be found in the article by Khanarian et al.[5]

## Electro-Optic Constant Measurement

The electro-optic or Pockels constant, r, is related to the bulk second order $\chi^{(2)}$ property as follows:

$$r = -2\chi^{(2)}/n^2\varepsilon \qquad (3)$$

The Pockels constant is a device figure of merit and is directly related to the change in the index of refraction which drives the device:

$$\Delta n = n^3 rV/W \qquad (4)$$

It thus becomes important to be able to measure this constant directly. After a film is poled, its r constant is measured by reflecting a laser beam off

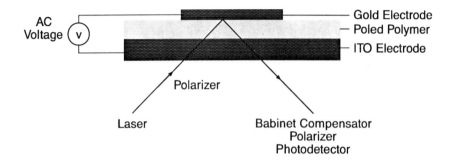

Figure 6    Measurement of r by Reflection

the top gold electrode as shown in Figure 6 while an a.c. signal is applied across the electrodes. The polarization of the input beam is at 45° and, due to the electro-optic effect, the polarization is rotated at the output. Through the use of a Soliel-Babinet compensator and an analyzer, this rotation is measured and, consequently, the r constant can be calculated. Figure 7 shows the r constant of the polymer HCC-1232 as a function of poling voltage.

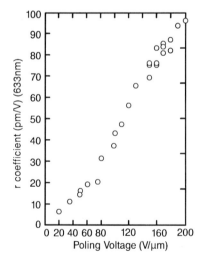

Figure 7    Polymer r Constant Versus Poling Voltage

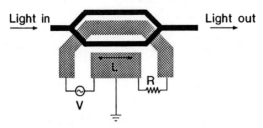

- Voltage varies along the length
  (L$\gtrsim \lambda$ electrical)
- Important device parameters:
  - Voltage
  - Drive power
  - Speed

Figure 8    Mach-Zehnder E-O Modulator

4 DEVICES

Organic devices are basically of two types: electro-optic and all optical. The former types control light with electronics and are based on the second order susceptibility $\chi^{(2)}$. The latter control light with light and are based on the third order effect $\chi^{(3)}$. This discussion will be limited to electro-optic devices such as the Mach-Zehnder modulator depicted in Figure 8. The device functions as follows. Light enters through the waveguide and is split into two equal arms, travels the length of the arms, and is re-combined at the other end. Thus you have light in and light out essentially unchanged. If electrodes are placed on one arm and, if these arms are constructed out of non-linear (electro-optic) material, the index of refraction will change as a function of the applied field. When this occurs, the speed of light, in this arm relative to the other, is changed. If there is sufficient voltage to produce a $\pi$ phase shift, the light beams interfere

|  | LiNbO$_3$ Modulator | | Polymer Modulator | |
| --- | --- | --- | --- | --- |
|  | Conventional | Aperiodic Structure | Current | Projected |
| Switching voltage (V) | 3.5 | 10.5 | 1.3 | 0.7 |
| Active length (mm) | 7.5 | 2.5 | 27 | 27 |
| Overall device length (cm) | 5 | 5 | 5.5 | 5.5 |
| Power (W) | 0.6 | 5.0 | 0.03 | 0.008 |
| Maximum freq. (GHz) | 8 | 24 | No limit other than electrical sources[1] | |
| Linear E-O coeff. (pm/V) | 31 | 31 | 35 | 70 |

(1) • Conventional electrodes: 50 GHz
• Super conducting electrode: ~500 GHz

Figure 9    Modulator Performance Comparisons

destructively and cancel each other. Thus, light in and no light out. One can also continually modulate the voltage to produce varying intensities. This very fast E-O switch can be used in such areas as telecommunications. Important device parameters to consider are: voltage, drive power, and speed. Performance comparisons of inorganic versus organic based M-Z devices are listed in Figure 9.

Based on an r coefficient of 31 pm/V, the lithium niobate modulator operates at 3.5 V with a power consumption of 0.6 W. The maximum frequency of operation is 8 GHz. Modifications to this device are shown in the second column and the operation frequency was raised to 24 GHz. However, the price was paid in operating voltage (10.5 V) and power consumption (5 W). On the other hand, a longer organic modulator, with a similar r value, operates at 1.3 V with a power consumption of only 0.03 W. Moreover, this device is not limited in frequency as is the inorganic device because there is no velocity mismatch between the electrical and optical signals. In order to achieve high frequencies in these devices, it is necessary that the light and electrical signals travel at the same speed through the device so that the light always interacts with the E field. This is the case for the organic, where the velocities are matched, but not for the inorganics. The polymer refractive index (light velocity) is 1.63 at optical frequencies and 1.73 at 20 GHz. For lithium niobate, the corresponding figures are 2.21 and 5.57, respectively. Thus, the latter device will be limited in frequency. A projected polymer device, with an r constant of 70, shows that switching voltage and drive power can be lowered even further.

## 5 SUMMARY

The presence of a non-linear optical response in a medium leads to a large number of technologically important processes and phenomena over a wide frequency range.

Organic polymers offer a series of significant property advantages over inorganic and organic crystals. Synthesis replaces difficult crystal growth techniques and can tailor the critical secondary properties necessary to achieve commercially viable materials. Poling controls the structure and results in high figures of merit. Polymers can be manufactured into a wide variety of shapes and forms, permitting unique device concepts not previously achievable with inorganic systems.

## 6 ACKNOWLEDGEMENTS

The authors gladly acknowledge the valuable technical guidance given to our effort by Professor A.F. Garito, University of Pennsylvania, and Drs. R. Lytel and G.F. Lipscomb, of the Lockheed Missiles and Space Company. The support, guidance and encouragement of Drs. E. Sharp and W. Elser of the Center for Night Vision and Electro-Optics, Drs. F. Patten and A. Yang of the Defense Advanced Research Projects Agency, and Dr. D. Ulrich of the Air Force Office of Scientific Research are gratefully acknowledged.

## 7 REFERENCES

1. G.J. Bjorklund, G. Carter, A.F. Garito, R.S. Lytel, G.R. Meredith, P. Prasad, J. Stamatoff, and M. Thakur, Appl. Opt., 1987, 26, 227; S.J. Lalama and A.F. Garito, Phys. Rev. A, 1979, 20, 1179; D.J. Williams, ed., "ACS Symposium Series", 233, American Chemical Society, Washington, D.C., 1983.

2. D. Chemla and J. Zyss, eds., "Nonlinear Optical Properties of Organic Molecules and Crystals", Vols. 1 and 2, Academic Press, Orlando, 1986.

3. C.S. Willand, S.E. Feth, M. Scozzafava, D.J. Williams, G.D. Green, J.I. Weinshenk, H.K. Hall, and J.E. Mulvaney, "Electric-Field Poling of Nonlinear Optical Polymers;" K.D. Singer, J.E. Sohn, and M.G. Kuzyk, "Orientationally Ordered Electro-Optic Materials," Proceedings of the ACS Symposium on Electro-Active Polymers, Plenum Press, New York, 1988.

4. Subsequent studies, with purer monomers, has resulted in the preparation of higher Tg polymers. However, the relationships still remain the same. (H.A. Goldberg, et al., MRS Proceedings, 1989, to be published).

5. G. Khanarian, T. Che, R. DeMartino, D. Haas, T. Leslie, H. Man, M. Sansone, J. Stamatoff, C. Teng, and H. Yoon, Proc. SPIE, 1988, 824, 72.

# Organic Materials for Nonlinear Optical Applications

D. Broussoux, P. Le Barny, J. P. Pocholle, and P. Robin

CENTRAL RESEARCH LABORATORY, THOMSON-CSF, 91404 ORSAY CEDEX, FRANCE

## 1 INTRODUCTION

Organic materials show attractive optical properties for nonlinear optical applications. Among all the forms in which these materials can be obtained, polymers are the most interesting for integrated devices. Indeed polymers can be deposited by using processes completly compatible with the silicon technologies. Their optical properties have been extensively studied and show great potential for use in electrooptic or frequency doubling devices.

But to achieve such devices, the long term operational stability of polymers has to be enhanced by blocking the mobility of the molecules. Different routes have been tried to reach this goal and the best one is to crosslink the molecules carrying the nonlinear group. Moreover, the trade off between low insertion loss, low absorption loss and high non linear properties has to be found depending on the applications. All these points will be discussed in this paper and illustrated by experimental results, and the critical modelisation. Then, in the last part of this paper, the reasonable limits of performance which polymeric devices could reach will be discussed.

## 2 MATERIALS FOR QUADRATIC NONLINEAR EFFECTS

To obtain a useful material possessing a large second order nonlinear susceptibility tensor $\chi^{(2)}$ requires the use of molecules with large microscopic second order nonlinear hyperpolarizability tensor ß organised in such a way that the resulting system has no centre of symmetry and an optimized constructive additivity of the molecular hyperpolarizabilities is achieved. In addition, the ordered structure thus obtained must not exhibit a loss of nonlinear optical properties with time. The nonlinear optical (NLO) active moieties which have been synthesized so far derive from the donor-$\pi$ system-acceptor molecular concept (Figure 1).

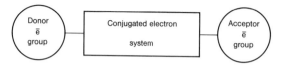

Figure 1  Basic molecular structure for quadratic nonlinear effects

Successful approaches to increasing the hyperpolarizabilities of NLO moieties have been achieved by using strong electron-donor and electron-acceptor groups and by increasing the conjugation length[1]. There is a strong correlation between enhanced non linearity and cut off wavelength : chromophores that absorb in the visible generally possess molecular hyperpolarizabilities much larger than those that are transparent[2-3]. So, depending on the used input laser wavelength and the type of application considered (electro-optic modulation or frequency doubling), a compromise has to be found between transparency and non linearity.

Ordered structures of NLO molecules, suitable for quadratic effects can be obtained from : single crystals, Langmuir-Blodgett (LB) films and polymer films.

Very large nonlinear coefficients have been obtained with single crystal organic materials[2]. But non-centrosymmetric single crystals are not common by nature; their growth and shaping are difficult and expensive. On the other hand many challenges remain in their application in guided wave structures. The attraction of the LB technique is that it is possible to build up supermolecular structures having a high degree of order in a direction normal to the substrate, by adding monolayers at a time. LB films exhibit interesting second order nonlinear effects, but again their use in guided wave devices seems questionable at the moment, since a partial regular structure within the plane of the layer induces optical losses by scattering and to obtain thick films, of the order of 1 micron thickness, requires more than 300 monolayers to be deposited.

Polymeric materials are emerging as a promising class of NLO materials because they exhibit the avantages of organic materials without having the disavantages of single crystals, they are compatible with semiconductor technology and they are particularly attractive for getting waveguides since they can be easily formed into thin low cost films. Three classes of quadratic NLO polymers are currently under investigation : liquid crystalline polymers (LCPs), ferro-electric polymers and amorphous polymers.

Polymeric films of sufficient thickness for single mode waveguiding are deposited onto a substrate via spincoating or dipping techniques. The obtained thickness depends on the viscosity of the solution, on the spinning or dipping speed, and on the type of substrate. After the deposition process, the NLO moieties inside the polymeric film are unoriented and the system is centrosymmetric. In order to obtain a non-centrosymmetric material, a DC electric field must be applied at a temperature which allows free orientation of the NLO groups (generally above the Tg of the material). This field acts on the ground state dipole moment of the NLO moieties causing them to align preferentially with the field, hence destroying the centrosymmetry of the film. Finally the obtained alignment is frozen by cooling down the film at room temperature before removing the electric field. The poling provides the alignment predicted by the Boltzmann's distribution law and imparts a $C_{\infty v}$ symmetry to the film (Figure 2).

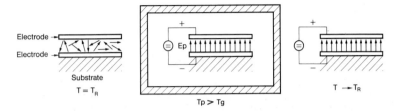

Figure 2    Electric field induced orientation of NLO moieties in a polymer
Tp : poling temperature   Ep : poling field   $T_R$ : room temperature

Liquid crystalline polymers

Side chain LCPs are expected to provide a higher degree of polar order than amorphous polymers. A lot of synthetic work has been done in this field[4-10] but, except for the earlier study of a molecularly doped LCP by Meredith et al.[11] no SHG experiments or electro-optic measurements using LCPs have been published so far. Very likely this situation is due to :
- the low Tg of the side chain LCPs (lower than 50°C in most cases) allowing a fast reorientation of the NLO moieties to take place and hence restoring the natural centrosymmetry.
- the difficulties to pole side chain LCPs. It appeared that the best way to get a non-centrosymmetric NLO side chain LCP is first to induce a homeotropic orientation by applying an AC electric field of about 20 V μm$^{-1}$ and a few kHz when the film is slowly cooling down from the isotropic phase and then to apply a strong DC electric field (about 80 V μm$^{-1}$) just below Tg over 24 hours[9].

Ferroelectric polymers

Two types of ferroelectric copolymers are currently studied : copoly(vinylidene fluoride - trifluoroethylene) [P(VDF-TrFE)], and copoly(vinylidene cyanide-vinyl acetate [P(VDCN-VAc)]. These two polymers are quite different in nature; P(PVF-TrFE) is a random semi-crystalline copolymer whereas P(VDCN-VAc) is an alternating amorphous copolymer (Figures 3 and 4).

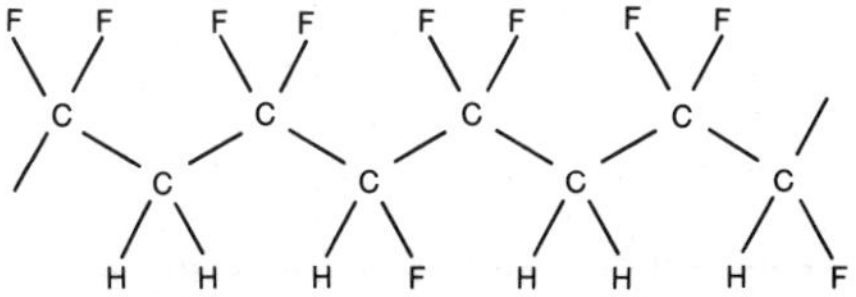

Figure 3    Chemical structure of P(VDF-TrFE)

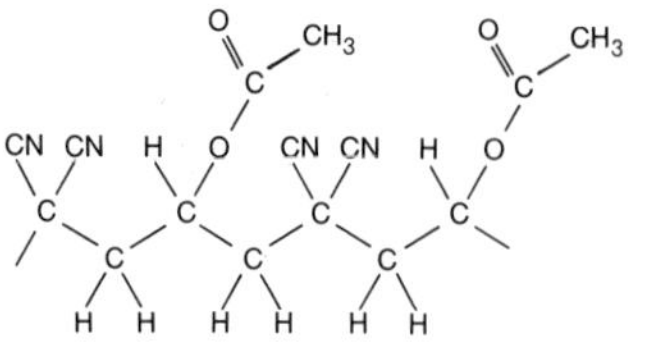

Figure 4   Chemical structure of P(VDCN-VAc)

In addition to their pyroelectric, piezoelectric and ferroelectric properties, poled P(VDF-TrFE) and P(VDCN-VAc) have been recognized as nonlinear optical materials[12-16]. Due to the weak hyperpolarizability of the C-F and C-C≡N bonds, the optical second harmonic coefficients $d_{33}$ exhibited by the two previous polymers is small. In Table 1 are summarized the main properties of P(VDCN-VAc) and a P (VDF-TrFE) having a trifluoroethylene content of 25 mole %[16]. Quadratic optical properties of P(VDF-TrFE) and P(VDCN-VAc) can be improved by dissolving NLO guest molecules in the ferroelectric host copolymers (guest-host-systems[17-18]). In the case of P(VDF-TrFE), guest molecules are located in the amorphous regions of the polymeric host. Solubility of guest molecules in ferroelectric copolymers is limited and the second harmonic coefficient of these guest-host systems at 1.064 μm was found to be few pm V$^{-1}$ [18]. Finally, due to their semi crystalline structure P(VDF-TrFE) copolymers are inclined to scatter light, preventing them from use in waveguided applications. This problem can be overcome by using P(VDCN-VAc) instead.

Table I   Comparative values for the copolymers P(VDF-TrFE) and P(VDCN-VAc)

|  | P(VDF-TrFE) 75/25 | P(VDCN-VAc) |
|---|---|---|
| Optical refractive index at λ = 0.63 μm | 1.42 | 1.47 |
| $d_{33}$ at 1.06 μm | 0.6 pmV$^{-1}$ | 0.4 pmV$^{-1}$ |
| Curie temperature | 125°C | 180°C |
| Tg | -30°C | 180°C |
| Useful temperature | up to 100°C | up to 140°C |

Amorphous polymers

At the moment, amorphous polymers seem to be the most promising polymeric material for quadratic NLO applications. During the last few years, the conception of amorphous polymeric materials has moved from molecularly doped polymers to covalently functionalized polymers and very recently to cross-linked polymers.

Doped amorphous polymers. Williams has first suggested the possibility of second order processes in orientationally ordered glasses[19] and this has been demonstrated by Singer and co-workers[20-24]. They have mainly studied the NLO properties of polymethyl-methacrylate (PMMA) doped with 4-[ethyl (2-hydroxyethyl)amino]-4'-nitroazobenzene, an azo dye also called Disperse Red 1 (Figure 5).

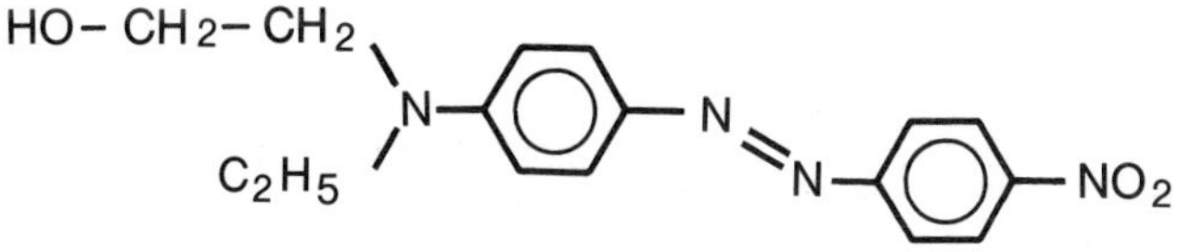

$$HO-CH_2-CH_2 \diagdown N - \bigcirc - N = N - \bigcirc - NO_2 \diagup C_2H_5$$

Figure 5   Chemical structure of Disperse Red 1

From simple thermodynamic considerations based on non interacting molecular dipoles, and taking into account the effect of local fields, it is possible to estimate the magnitude of the largest second harmonic coefficient component $d_{33}$ of such a poled guest-host system

$$d_{33}(-2\omega,\omega,\omega) = N\, f(2\omega)\, f^2(\omega) \times \frac{\varepsilon_r\,(n^2+2)}{(n^2+2\varepsilon)} \times \frac{\mu_0 E_p}{5\,kT} \times \beta_{zzz} \qquad (1)$$

where :

N   is the number of NLO molecules per unit volume
$f(2\omega)$ and $f(\omega)$ are local field factors given by :

$$f(\omega) = \frac{n^2(\omega)+2}{3} \quad \text{and} \quad f(2\omega) = \frac{n^2(2\omega)+2}{3} \qquad (2)$$

$\beta_{zzz}$ is the molecular hyperpolarizability component along the molecular Z axis
$\varepsilon_r$ is the static dielectric constant
$\mu_0$ is the ground state dipole moment of the NLO molecules
$E_p$ is the poling field
$n$ is the optical index of refraction

Singer et al. have proved the validity of the model by checking the linear dependance of $d_{33}$ versus poling field Ep and number density N. $d_{33}$'s values as high as 31 pm/V have been obtained just after poling at a incident wavelength of 1.58 μm from a dicyanovinylazo dye (DCV) dissolved in PMMA[25] (N = 2.3 $10^{20}/cm^3$) but guest-host systems suffer from the relatively low solubility of the dye in the host and from the relaxation of the orientational ordering obtained by poling, which occurs at room temperature (Figure 6). These relaxations may result from a dipole - dipole induced reorientation and a diffusion of the dye in the polymer.

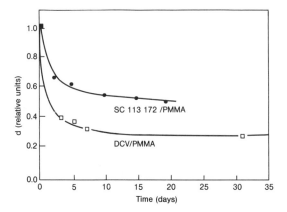

Figure 6  Decay of second-harmonic coefficient of SC 113 172/PMMA[26] (a) and DCV/PMMA[25] (b)

This behaviour has been confirmed by Le Barny et al[26] in a guest host system composed of a PMMA matrix containing 10 weight % of SC 113 172, an ICI, azo dye, whose structure is described on Figure 7.

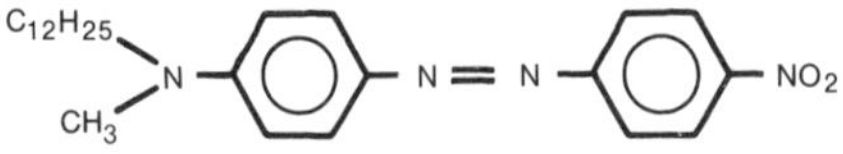

Figure 7  Chemical structure of SC 113 172

Mechanisms of the relaxation processes are not well understood. Local free volume and/or local mobility in the polymer matrix seem to play an important role[27].
This assumption is supported by the following experimental observations :
- The rate of decay for a given polymer host increases with decreasing dopant size.
- The temporal stability depends on the nature of the polymer host. For example, for a given dopant, SHG signal decreases more rapidly in PMMA than in polystyrene (PS) even though both polymers have practically the same Tg (about 100°C). This behaviour is explained by the existence of a broad secondary transition for PMMA at higher temperatures (35 to 50°C) than for PS (- 50°C to - 30°C)[28].
- Physical ageing of the films during annealing at room temperature after poling leads to an improved temporal stability of the SHG signal. This result is consistent with a decrease in chain segment mobility of the polymer host. Hampsch et al[27] have tried to fit their experimental results to the Williams-Watts stretched exponential :

Organic Materials for Nonlinear Optical Applications 243

$$y = e^{(-t/\tau)b} \quad (3)$$

where y represents the normalized SHG intensity, $\tau$ is a characteristic relaxation time and b reflects the breadth of distribution of relaxation times or measures the cooperativity of the relaxation processes[29]. It appears that the model is insufficient to describe the behaviour of all the guest-host systems studied.

Covalently functionalized amorphous polymers. A natural step towards highly efficient and stable NLO polymers was the synthesis of covalently functionalized amorphous polymers. By linking directly the dye to the polymer backbone, it is possible to achieve high chromophore densities without phase separation and to increase the temporal stability of the poled material. Covalently functionalized amorphous polymers can be obtained either by copolymerization of a monomer bearing the chromophore unit via a flexible spacer group with a comonomer contributing to the amorphous character of the final copolymer[30-34] or by modifying a reactive polymer such as poly (p-hydroxystyrene)[35-37], poly [methyl-methacrylate-co-(N-ethyl anilino) ethyl methacrylate][38-39], poly (hydrogen methyl siloxane) or poly (allylamine hydrochloride)[41]. Chromophore functionalization levels from 0 to 100% can thus be achieved at will.

Studies of the nonlinear coefficient as a function of the dye content in two families of copolymers[36-30] (Figure 8) have shown a fairly linear dependance (Figure 9).

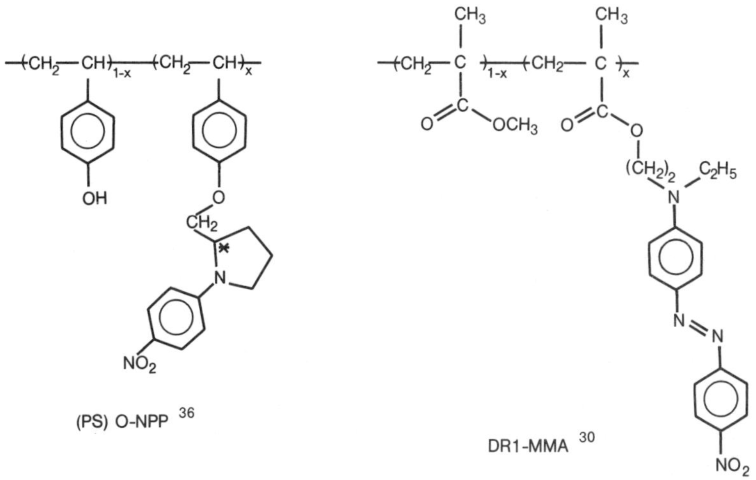

Figure 8   Chemical formula of the two families of copolymers used to study the behaviour of $d_{33}$ as a function of x

This result indicates that the noninteracting oriented molecular gas model also works for copolymers having at least 48 % chromophore functionalization level.

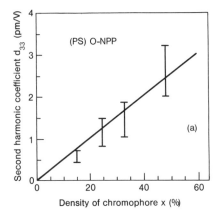 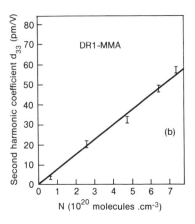

Figure 9   Second harmonic coefficients $d_{33}$ of (PS) O-NPP films poled at 30 V $\mu m^{-1}$ as a function of chromophore functionalization level for 1.064 μm incident radiation (a) and $d_{33}$ of DR1-MMA films poled at 85 V $\mu m^{-1}$ as a function of x for 1.064 μm incident radiation (b).

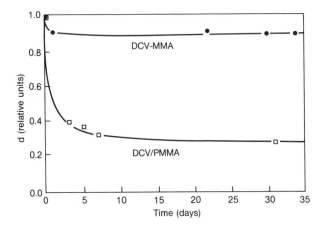

Figure 10   Decay of second-harmonic coefficient of DCV-MMA and DCV/PMMA corona-poled films. The chemical structure of DCV-MMA copolymer is depicted in Figure 11.

As expected, an improvement of the stability of NLO properties of covalently functionalized polymer (DCV-MMA for example) compared to guest-host polymer systems (DCV/PMMA), is observed[38] in Figure 10.

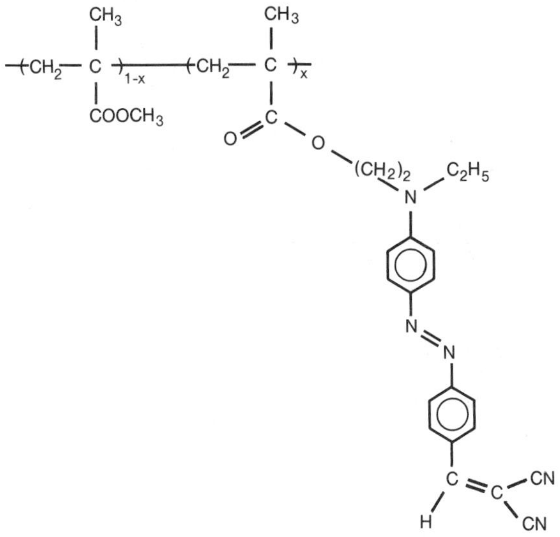

Figure 11   Chemical formula of DCV-MMA copolymer[38]

Recently, the temporal stability of the orientational ordering obtained by poling covalently functionalized amorphous polymers has been investigated mainly by using (PS) O - NPP copolymers (Figure 8). Again the relaxation processes are not well identified, but in Figure 12 the $d_{33}$ decay can be fitted to a two-exponential model[36-37] written as :

$$d_{33} = \underbrace{A e^{-t/\tau_1}}_{\text{Fast component}} + \underbrace{B e^{-t/\tau_2}}_{\text{Slow component}} \qquad (4)$$

$\tau_1$ and $\tau_2$ are respectively the short-term and the long-term SHG decay lifetimes. The maximum derived $\tau_2$ value is as large as 313 days (15 % functionalization level) but $\tau_2$ is dependent on the functionalization level and decreases as the chromophore content increases (Figure 13). This behaviour could be explained by the decrease in hydroxy phenyl-hydroxy phenyl hydrogen bonding as the functionalization level increases. On the other hand, it has been found that the amplitude of the short term process is connected to the presence of local free volume allowing the local movement of chromophores.

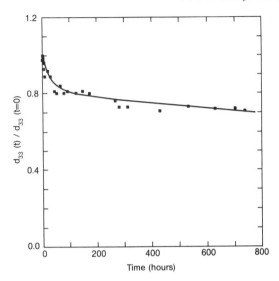

Figure 12  Temporal behaviour of the second harmonic coefficient of an annealed (PS)O-NPP film (25 % functionalization level) at room temperature. The solid line shows a two exponential fit (eq. (4)) to the data points.

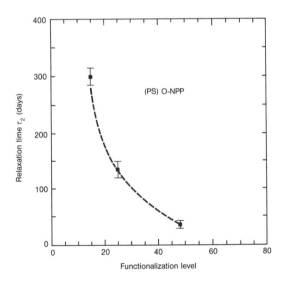

Figure 13  Dependence of the second harmonic coefficient exponential lifetime $\tau_2$ on functionalization level for (PS)O-NPP[36]

This is supported by the following observations : elimination of residual solvent in the polymer by annealing just after spin coating, greatly slows down the fast $d_{33}$ decay and as the ageing tempeature gets higher, decay becomes faster, indicating the greater ease of chromophore movement[42] (Figure 14).

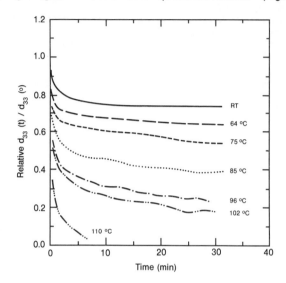

Figure 14 $d_{33}$ decay curves of (PS)O-NPP (48 % functionalization level) at various ageing temperatures

$\tau_2$ can be regarded as originating from the reorganization of entire polymer chains which takes a much longer time than does the local movement.

Until now, most of the studied copolymers have chromophores in their structure with high ß µ values. Recently, at Thomson-CSF, we have synthesized two new copolymers PLBP 73 and PLBP 80 respectively designed for electrooptic modulation with an input wavelength of 630 nm and second harmonic generation of a Nd-YAG laser ($\lambda$ = 1064 nm). The main properties of PLBP 73 and PLBP 80 are summarized in Table 2.

Table 2 Non linear properties of copolymers PLBP 73 and PLBP 80

| Index | Dye content (%) | Tg (°C) | $\lambda_{max}$ (nm) | $d_{33}$ (pm/V) | $n_3$ at 0.63 µm |
|---|---|---|---|---|---|
| PLBP 73 | 35 | 110 | 394 ($CHCl_3$) | 13 | 1.59 |
| PLBP 80 | 45 | 112 | 348 ($CHCl_3$) | 12 | 1.59 |

Cross-linked polymers. Development of NLO polymers is strongly connected to the possibility of obtaining a stable material.

Cross-linking studies, which appears at the moment as the ultimate possibility to achieve such a material are under way. Recently, three different approaches were considered :

- Immobilization of NLO molecules within a two-component optically transparent thermosetting epoxy[44] (EPO-TEK 301-2).
- Cross-linking of a X-ray resist bearing a NLO moiety[45].
- Chemical cross-linking of a tetrafunctional NLO molecules bearing cross-linkable functions[46].

In all cases cross-linking was achieved under electric field to obtain the desired alignment of NLO chromophores.

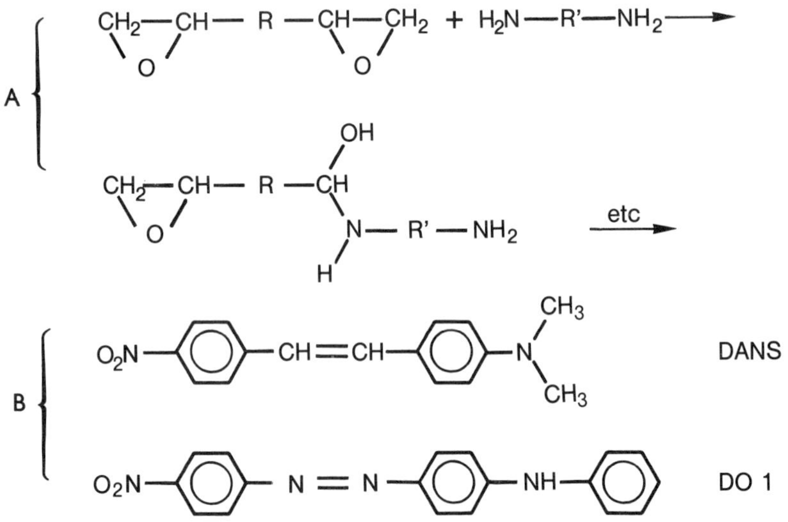

Figure 15  Chemical structure of the guest-host systems used by M.A. Hubbard et al[44] ; A = host : EPO-TEK 301-2 ; B = NLO molecules

The material obtained by the first approach exhibits a relaxation of its NLO properties. Again, the $d_{33}$ decay fits to a two-exponential model such as equation (4).

The two other approaches led to stable systems (Figure 18).

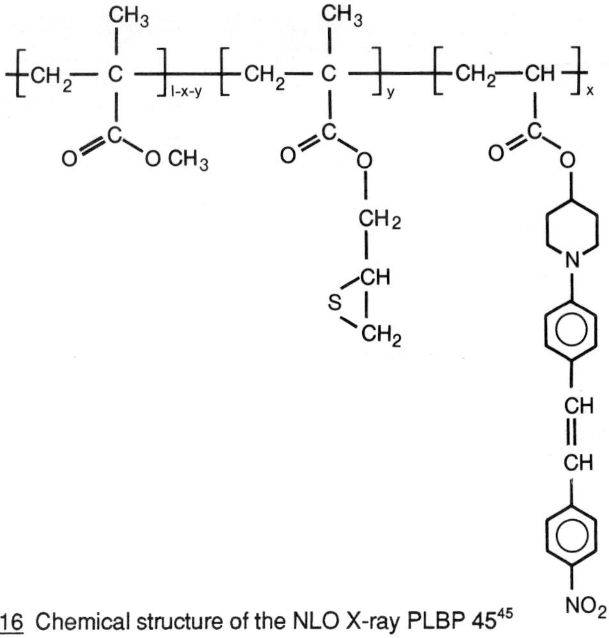

Organic Materials for Nonlinear Optical Applications 249

Figure 16  Chemical structure of the NLO X-ray PLBP 45[45]

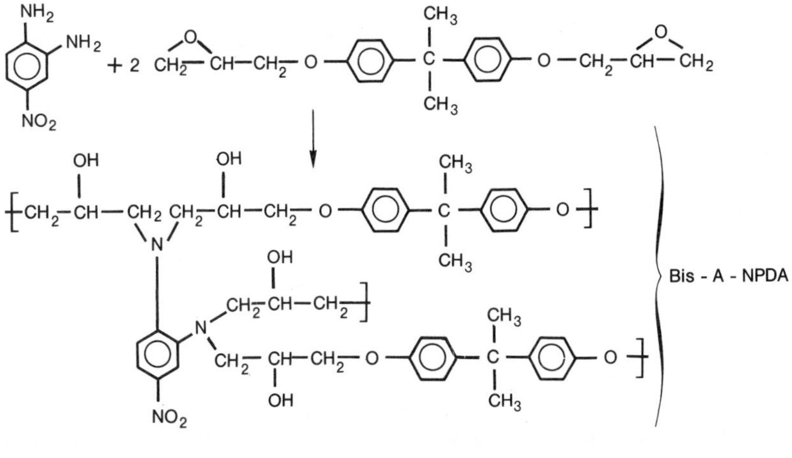

Figure 17  Chemical structure of the crosslinked system studied by M. Eich et al[46]

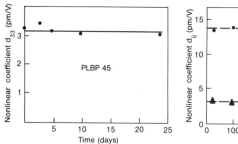
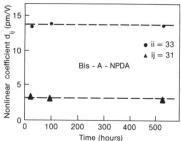

**Figure 18** Ageing of NLO properties of PLBP 45 and bis-A-NPDA at room temperature

Bis-A-NPDA showns no tendency of relaxation even at 85°C.
These significant results allow us to conclude that cross-linking seems to be the best way to obtain a stable NLO polymeric material. However, a lot of chemical process work has to be carried out to obtain a stable system with all the desired properties (efficiency, transparency...).

## 3 FREQUENCY DOUBLING

When a high intensity optical wave at $\omega$ frequency interacts with a non-centro-symmetrical material, a second order polarization is induced. That nonlinear polarization radiates at the double frequency of the incident beam. The created second harmonic intensity can be written in the parametric approximation as :

$$I(2\omega) = K \times I^2(\omega) \, e^{-(\alpha(\omega) + \alpha(2\omega)/2)l} \times 2 \, \frac{(\cosh(\Delta\alpha l) - \cos(\Delta k l)}{(\Delta k^2 + \Delta\alpha^2)} \quad (5)$$

with $K = 2\omega^2/\varepsilon_0 c^3 \times$ Figure of merit (F) (6)

and $F = d^2_{eff} / n(2\omega) \, n^2(\omega)$ (7)

$d_{eff}$ is the effective non linear coefficient, $n(\omega)$ and $n(2\omega)$ are respectively the indices at $\omega$ and $2\omega$, $\alpha(\omega)$ and $\alpha(2\omega)$ are respectively the absorption coefficient at $\omega$ and $2\omega$, l is the interaction length, $I(\omega)$ is the pump intensity, c and $\varepsilon_0$ are respectively the light velocity and the dielectric constant of the vacuum.

In expression 5 ($\Delta k.l$) represents the term of phase-matching between the pump wave at $\omega$ and the second harmonic wave at $2\omega$ as :

$$\Delta k = k(2\omega) - 2k(\omega) \, ; \, |k(\omega)| = 2\pi \, n(\omega)/\lambda(\omega) \, ; \, |k(2\omega)| = 2\pi \, n(2\omega)/\lambda(2\omega) \quad (8)$$

and $\Delta\alpha$ represents the difference of the absorption at the frequencies $\omega$ and $2\omega$ as :

*Organic Materials for Nonlinear Optical Applications*

$$\Delta\alpha = \alpha(\omega) - \alpha(2\omega)/2 \tag{9}$$

At an operating frequency without optical losses, $I(2\omega)$ can be written as :

$$I(2\omega) = K \times (I(\omega))^2 \times l^2 \times \text{sinc}^2 (\Delta kl/2) \tag{10}$$

If $\Delta k = 0$, the conversion efficiency decreases with the interaction length. If the conversion factor is defined as,

$$C = I(2\omega)/I(\omega) = K \cdot I(\omega) \, l^2 \, \text{sinc}^2 (\Delta kl/2) \tag{11}$$

we can notice that C is proportional to $I(\omega)$ and is a maximum when $\Delta kl/2 = \pi$ for a critical interaction length called $l_{co}$, the coherence length :

$$l_{co} = \lambda(\omega)/4 \, (n(2\omega) - n(\omega)) \tag{12}$$

$l_{co}$ becomes   leading to a high conversion values of C in the case of :

$$n(2\omega) = n(\omega) \tag{13}$$

That condition can be obtained by exploiting the dispersion of the birefringence of the material.

## 4 ELECTROOPTIC MODULATION

The variation of the coefficient $1/n^2_{ij}$ of the index ellipsoid is :

$$\Delta \left(\frac{1}{n^2}\right)_{ij} = r_{ijk} E_k + s_{ijkl} E_k E_l \tag{14}$$

where $r_{ijk}$ and $s_{ijkl}$ are respectively the linear electrooptic (Pockels) coefficient and quadratic (Kerr) coefficient. $E_k$ is the $k^{th}$ electric field's component applied to the structure. The elements of the tensor [s] being smaller than the elements of the tensor [r], the quadratic effect can be neglected.

The electrooptic coefficient can be related to the second order susceptibility by :

$$\chi_{ijk}^{(2)} (-\omega,\omega,0) = -1/2 \, [n_{ij}^2 (\omega) \, n_{ij}^2 (0)] \cdot r_{ijk} (-\omega,\omega,0) \tag{15}$$

This expression shows that the electrooptic coefficients are proportional to the second order susceptibility. By contracting the indexes following the Voigt notation due to the conditions of symmetry, the tensor [r] can be written in a $\infty$ mm point group symmetry as :

$$[r] = \begin{bmatrix} 0 & 0 & r_{13} \\ 0 & 0 & r_{23} \\ 0 & 0 & r_{33} \\ 0 & r_{42} & 0 \\ r_{51} & 0 & 0 \\ 0 & 0 & 0 \end{bmatrix}$$

If we refer to Figure 19 an optical wave will see a phase shift induced by the applied electric field $E_3$ :

$$d\varphi = \frac{\pi}{\lambda} n_e^3 [r_{33} - (n_o^3/n_e^3) r_{13}] E_3 \qquad (16)$$

where $r_{33}$ and $r_{13}$ are the electrooptic coefficients for the modes TM and TE. $n_2$ is equal to $n_o$, the ordinary index and $n_3$ is equal to $n_e$, the extraordinary index.

From this expression, we can deduce the voltage required for a total extinction :

$$V_\pi = \frac{\lambda}{n_e^3 [r_{33} - (n_o^3/n_e^3) r_{13}]} \times \frac{e}{l} \qquad (17)$$

If we assume that $(n_o^3/n_e^3) r_{13} \ll r_{33}$, we obtain :

$$V_\pi \sim \frac{\lambda}{n_e^3 r_{33}} \times \frac{e}{l} \qquad (18)$$

With this relation, we can see that it is necessary to minimize the ratio e/l, i.e. to have l >> e. This is the case in a guided wave structure where e/l is typically equal to $10^{-4}$.

In general an electrooptic modulator defined as a waveguide structure can be characterized by four principal criteria ; modulating voltage, bandwidth, power requirements and propagation losses.

Modulating voltage

To calculate the modulating voltage, we have to define the waveguide parameters such as the thickness of the guide in order to get a monomode propagation and its length and width in order to get a large modulation efficiency.

Taking the $V_\pi$ expression, we introduce a confinement factor $\eta$ which is the overlap integral between the electrooptically active layer and the guided mode :

$$V_\pi = \frac{\lambda}{n_e^3 r_{33}} \times \frac{e}{l} \times \frac{1}{\eta} \qquad (19)$$

At $\lambda = 1.32$ µm with $r_{33} = 20$ pm/V, e = 2.5 µm, l = 1 cm and $\eta = 0.8$, we obtain $V_\pi = 6$ V. The modulation efficiency can also be defined as the number of degrees phase shift achieved with 1 V over 1 mm propagation length :

$$\varphi = \frac{180°}{V_\pi \times l} = 3°/V/mm \qquad (20)$$

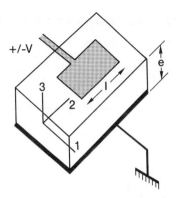

Figure 19 Structure of an electrooptic modulator

The latter value is close to 2.5°/V/mm for $LiNbO_3$ based devices.

Bandwidth

One of the advantage of an organic guide is that we can optimize the mismatch between the microwave phase velocity and the guided light velocity, respectively defined by $c/\sqrt{\varepsilon_e}$ and $c/n_e$, where $\varepsilon_e$ and $n_e$ are the effective dielectric constant and the optical index.

Without a buffer layer, we get:

$$n_e = \sqrt{\frac{1 + \varepsilon_r}{2}} = 1.414 \tag{21}$$

$n_o = 1.51$

and the bandwidth at 3 dB (for l = 1 cm) is:

$$B_{-3dB} = \frac{0.443 \, c}{l. \, |n_e - n_o|} = 138 \text{ GHz} \tag{22}$$

if we assume that all the propagating waves have no loss.

## Power requirements

A limitation can be the power required to get a π phase shift : with a modulator using an organic layer, we can expect to obtain a low power. For a modulation length of 2 cm and a $V_\pi$ = 3 V with electrodes calculated to obtain 50 Ω we obtain a power P = 180 mW. This value must be compared with those of $LiNbO_3$ equal to 300 mW under the same conditions. This outlines the potential advantage of organic materials for electrooptic modulation over more conventional material systems.

## Optical losses

The optical propagation losses are an important parameter to cope with for making high performance devices. In this field, $LiNbO_3$ has very low absorption losses (< 0.1 dB/cm) though for organic materials, we can expect to achieve losses < 0.5 dB/cm.

## 5 INTEGRATED OPTICAL DEVICES

Integrated Optical Devices (IOD) require 2D waveguides. Different methods can be used such as chemical etching, ionic diffusion, cross-linking (UV, X, e...) for getting Ve, ribbon or local variation of index. Figure 20 shows some of the different structures that are achievable in a 2D waveguide.

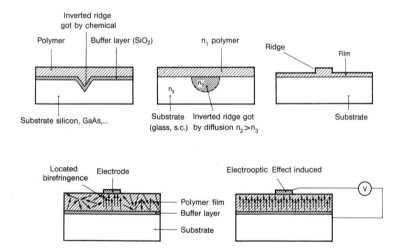

Figure 20 Different realizations of 2D optical structures

## Second harmonic generator (SHG)

Guided optics permit us to control exactly the effective index required for the propagation of an electromagnetic wave. Indeed the phase which characterizes the guided wave in a three layers structure depends on the conditions at each interface. Figure 21 shows the behaviour of the effective index relating to the optical mode versus the thickness of the active layer.

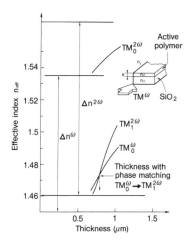

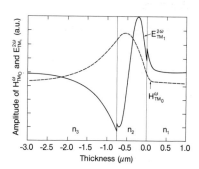

Figure 21 Effective index versus the thickness of the active layer

Figure 22 Amplitude distribution of magnetic and electric optical modes versus the thickness of the three layers structure

We can notice that the curve of the index's dispersion $TM_0(\omega)$ cuts the curve $TM_1(2\omega)$ for the thickness 0.76 µm. For this value the condition of phase matching as $n_{eff}(\omega) = n_{eff}(2\omega)$ is satisfied. But this condition is not sufficient for getting a high conversion efficiency and the value of the overlap integral must be as high as possible. Figure 22 shows the distribution of the amplitude of the modes $H_{TM_0}(\omega)$ and $E_{TM_1}(2\omega)$ calculated for a three layers'structure. The behaviours of these components are characteristic of a symmetric distribution and of an asymmetric distribution. Figure 23 shows the particularly low values of the overlap integral corresponding to the case described in Figure 21. But in this case, the values of the coherence length are particularly attractive. The other methods proposed by Bloembergen[47] realize an artificial phase matching by using an alternated index modulation along the propagating direction. The period is $\Lambda = 2ml_{co}$ with m = 1,3,5... and the relation between the wave vectors is $k_2 - 2k_1 - K = 0$ where K, is the wave vector associated with the distribution of the nonlinear coefficient, equal to $K = 2m\pi/\Lambda$. In polymers such modulations can be obtained by creating alternating regions poled up and poled down. That is equivalent to using a nonlinear material with a efficiency reduced by a factor $2/m\pi$. As reported in Figure 23 for the case 1, the case 2 represented in Figure 24 shows interesting values of the overlap integral while the values of the coherence length are particularly low. But its monotonous behaviour permits a larger flexibility for the realization of the optical structure.

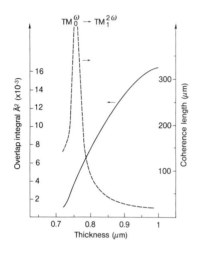 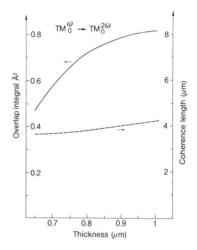

Figure 23 Overlap integral and coherence length versus the thickness using the indices dispersion

Figure 24 Overlap integral and ccherence length versus the thickness using a pseudo phase matching

Electrooptic modulator (EOM)

The achievement of optical components operating over the range [0.8 - 1.6 µm] and activated by an electric field requires the use of new materials with high electrooptic coefficients and low absorption losses. Polymers are sufficiently flexible to permit the realization of the trade off between high electrooptic efficiency and low absorption losses. Authors[48-49] have already reported large values of electrooptic coefficients (50 to 100 pm/V) but the values measured on optical structures are generally lower (5 to 20 pm/V). Indeed for using integrated waveguides low absorption losses are required, so such high values of $r_{33}$ obtained near the absorption band cannot be used because of high absorption losses associated with the resonnant effect. At Thomson-CSF a planar electrooptic modulator has been achieved by using a material PLBP 73 showing a $r_{33}$ = 12 pm/V at 0.6328 µm. Figure 25 shows the typical structure composed of three layers spin coated onto a conductive silicon substrate.

*Organic Materials for Nonlinear Optical Applications* 257

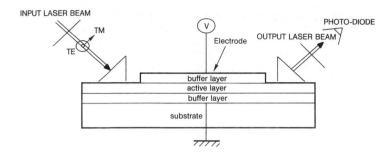

Figure 25  Structure of a planar electrooptic modulator used for measuring electrooptic coefficient and optical indices

This modulator operates by using the differential birefringence electrically induced between TE and TM which is described by the relation :

$$I = \sin^2 \Delta\varphi/2 \qquad (23)$$

where $\Delta\varphi$ is given by the relation (16). For characterizing the electrooptic coefficient the method used here measures $\partial I/\partial V$ as a function of $V$ by assuming :

$$r_{33} = r^0_{33} + \alpha V/e \qquad (24)$$

where V is the applied voltage, e the thickness and $\alpha$ a coefficient of proportionality related to $\chi^{(3)}$ and electrostriction phenomena.
Figure 26 shows the amplitude of the maximum which increases with the static applied voltage leading to a continuously varying $V_\pi$. To take into account this effect it is necessary to add the real $r_{33}$ described by the relation (24).
Figure 27 shows the calculated curve which fits rather well the experimental curve in Figure 26 by using the relation :

$$\partial I/\partial V = (r^0_{33} + 2\alpha V/E) B/2 \sin[2(r^0_{33} + \alpha V/e) BV] \qquad (25)$$

rather than the unmodified relation

$$\partial I/\partial V = r^0_{33} B/2 \sin[2 r^0_{33} BV] \qquad (26)$$

where $B = n^3 \pi L/3\lambda_1 \varepsilon$ \qquad (27)

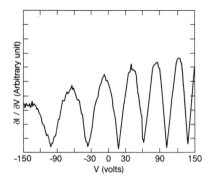
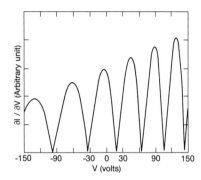

Figure 26 Measured values of ∂I/∂V versus the applied voltage V

Figure 27 Calculated values of ∂I/∂V versus the applied voltage V

This method allows us to obtain the true values of $V_\pi$ and $r_{33}$ but for increasing the accurancy of this method, it is necessary to have the exact values of the optical index at different wavelengths. It is difficult to achieve these values at large intervals of wavelength but the relation of Sellmeir on the dispersion allows an extension of the values of indices over a large interval :

$$n^2 = A + B / (1 - C/\lambda^2) \qquad (28)$$

in which we assume that $C = \lambda_o^2$ and A and B are constants depending on the material.

For PLBP 73, by taking A = 2.1698, B = 0.2182 and $C = \lambda_o^2$ with $\lambda_o$ = 384 nm, Figure 28 shows that the experimental results are fitted by the relation[28].

The waveguide losses of the waveguides were measured by probing the scattered light as a function of the distance along the guide. This was achieved by recording with a linear photodiodes array. Propagation losses of 3 dB/cm at 0.632 μm corresponding to a $V_\pi$ = 13 volts were measured on 1 cm long waveguides.

In the same way, another polymer PLBP 75, useful for near infrared applications leads to $V_\pi$ = 20 volts with a propagation loss of 0.6 dB/cm at 1.32 μm.

*Organic Materials for Nonlinear Optical Applications* 259

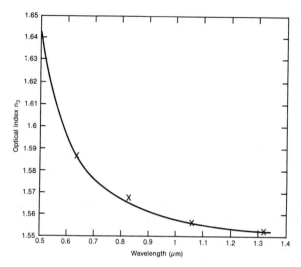

Figure 28  Experimental (x) index measurements for PLBP 73 reported on the calcultated curve using Sellmeir's relation

## 6 CONCLUSION

Today many applications can be projected using new stable crosslinked polymers. Simple deposition techniques on different substrates allow us to achieve optical guided structures, comparable to those achieved with $LiNbO_3$ and with semiconductors. So it will be possible to realize integrated couplers, interferometers, lenses... The good active properties of organic materials allow the realization of low loss electrooptic modulators with $V_\pi < 5$ V operating at high frequency (10-100 GHz) and second harmonic generators with doubling efficiency close to 10 % for doubling a semiconductor laser operating at 0.8 µm. Work engaged at Thomson-CSF in this field progresses in this direction and the first results obtained on materials have already shown that such devices could be achieved in the next five years, mainly in the fields of high density recording and high speed information processing.

### ACKNOWLEDGEMENTS

A part of this work is supported by EEC under Esprit 2284 and the other part of this work is supported by Ministere de la Recherche et de la Technologie.

## REFERENCES

1. K.D. Singer, J.E. Sohn, L.A. King, H.M. Gordon, H.E. Katz and C.W. Dirk, J. Opt. Soc. Am. B, 1989, 7, 1339.
2. D.S. Chemla and J. Zyss, eds., 'Nonlinear optical properties of organic molecules and crystals', Academic Press, New York, 1987.
3. J.L. Oudar, J. Chem. Phys., 1977, 67, 466.
4. A.C. Griffin, A.M. Bhatti and R.S.L. Hung, Proceedings of SPIE, 1987, 65, 682.
5. P. Le Barny, G. Ravaux, J.C. Dubois, J.P. Parneix, R. Njeumo, C. Legrand and A.M. Levelut, Proceedings of SPIE, 1987, 682, 56.
6. T.M. Leslie, R.N. De Martino, E.W. Choe, G. Khanarian, D. Haas, G. Nelson, J.B. Stalmatoff, D.E. Stuetz, C.C. Teng, H.N. Yoon, Mol. Cryst. Liq. Cryst., 1987, 153, 451.
7. A.C. Griffin, A.M. Bhatti and R.S.L. Hung, Mol. Cryst. Liq. Cryst., 1988, 155, 129.
8. C. Noel, C. Friedrich, V. Leonard, P. Le Barny, G. Ravaux and J.C. Dubois, Makromol. Symp., 1989, 24, 283.
9. R.N. De Martino, H.N. Yoon, J.B. Stamatoff, Eur. Patent 0271 730.
10. R.N. De Martino, Eur. Patent 0 294 706.
11. G.R. Meredith, J.G. Van Dusen and D.J. Williams, Macromolecules, 1982, 15, 1385.
12. J.H. Mc Fee, J.G. Berman and G.R. Crane, Ferroelectrics, 1972, 3, 305.
13. H. Sato and H. Gamo, Jpn J. Appl. Phys., 1986, 25, L990.
14. H. Sato, T. Yamamoto, I. Seo and H. Gamo, Optics Letters, 1987, 12, 579.
15. D. Broussoux, E. Chastaing, S. Esselin, P. Le Barny, P. Robin, Y. Bourbin, J.P. Pocholle and J. Raffy, Rev. Tech. Thomson-CSF, 1989, 20-21 (1) 151.
16. T. Kobayashi eds, 'Nonlinear optics of organics and semiconductors', Springer Verlag, 1989, 36, 126.
17. P. Pantelis, J.R. Hill, S.N. Oliver and G.J. Davies, Br Telecom. Technol. J., 1988, 6, 3.
18. J.R. Hill, P. Pantelis, P.L. Dunn, S.N. Oliver and G.J. Davies, Proceedings of 1st International Conference on Electrical Optical and Acoustic properties of Polymers, 1988.
19. D.J. Williams, Angew. Chem. Intl. Ed. Eng., 1984, 23, 690.
20. K.D. Singer, S.J. Lalama and J.E. Sohn, Proceedings of SPIE, 1985, 578, 168.
21. K.D. Singer, J.E. Sohn and S.J. Lalama, Appl. Phys. Lett., 1986, 49, 5, 248.
22. R.D. Small, K.D. Singer, J.E. Sohn, M.G. Kuzyk and S.J. Lalama, Proceedings of SPIE, 1986, 682, 160.
23. K.D. Singer, S.J. Lalama, J.E. Sohn and R.D. Small "Nonlinear optical properties of organic molecules and crystals", D.S. Chemla and J. Zyss eds, Academic Press Inc., 1987, 1, 437.
24. K.D. Singer, M.G. Kuzyk and J.E. Sohn, J. Opt. Soc. Am. B, 1987, 4-6, 968.
25. K.D. Singer, M.G. Kuzyk, W.R. Holland, J.E. Sohn, S.J. Lalama, R.B. Comizzoli, H.E. Katz and M.L. Schilling, Appl. Phys. Lett., 1988, 53, 19, 1800.
26. P. Le Barny, S. Esselin, D. Broussoux, J. Raffy, J.P. Pocholle, Proceedings of SPIE, 1987, 864, 2.

27. H.L. Hampsch, J. Yang, G.K. Wong and J.M. Torkelson, Polymer Commun., 1989, 30, 40.
28. H.L. Hampsch, J. Yang, G.K. Wong and J.M. Torkelson, Macromolecules, 1988 21, 526.
29. I.M. Hodge, Macromolecules, 1983, 16, 898.
30. S. Esselin, P. Le Barny, P. Robin, D. Broussoux, J.C. Dubois, J. Raffy, J.P. Pocholle, Proceedings of SPIE, 1988, 971, 120.
31. R.N. De Martino, H.N. Yoon, US Patent 4,801,670.
32. R.N. De Martino, Eur. Pat. 0 316 662.
33. C.E. Won, Eur. Pat. 0 306 893
34. R.N. De Martino, Eur. Pat. 0 312 856
35. C. Ye, T.J. Marks, J. Yang and G.K. Wong, Macromolecules, 1987, 20, 2322.
36. C. Ye, N. Minami, T.J. Marks, J. Yang and G.K. Wong, Mat. Res. Soc. Symp. Proc., 1988, 109, 263.
37. D. Li, N. Minami, M.A. Ratner, C. Ye and T.J. Marks, Synthetic Metals, 1989, 28, D 585.
38. K.D. Singer, M.G. Kuzyk, W.R. Holland, J.E. Sohn, S.J. Lalama, R.B. Comizzoli, H.E. Katz and M.L. Schilling, Appl. Phys. Lett., 1988, 53, 19, 1800.
39. M.L. Schilling, H.E. Katz and D.I. Cox, J. Org. Chem., 1988, 53, 5538.
40. T.M. Leslie, US Patent 4 801 659.
41. M. Eich, A. Sen, H. Looser, G.C. Bjorklund, J.D. Swalen, R. Twieg and D.Y. Yoon, J. Appl. Phys., 1989, 66, 6, 2259.
42. N. Minami, C. Ye, T.J. Marks and G.K. Wong, Polymer Preprints, Japan, SPSJ 38th Symposium on macromolecules, 1989, 38, 3K02.
43. P. Le Barny, D. Broussoux, French Patent 89 11327.
44. M.A. Hubbard, J.J. Marks, J. Yang and G.K. Wong, Chemistry of Materials, 1989, 1, 2, 167.
45. P. Le Barny, D. Broussoux, S. Esselin, J.P. Pocholle, J. Raffy, French Patent, 88 05790.
46. M. Eich, B. Reck, D.Y. Yoon, C.G. Wilson and G.C. Bjorklund, J. Appl. Phys. 1989, 66, 7, 3241.
47. N. Bloembergen and A.J. Sievers, Appl. Phys. Lett., 1970, 17, 11, 483.
48. R. Lytel, G.F. Lipscomb, M. Stiller, J.I. Thakckara and A.J. Ticknor, Proceedings of SPIE, 1988, 971, 218.
49. G.R. Möhlmann, C.P.J.M. Van der Voast, R.A. Huigts and C.T.J. Wreesmann, Proceedings of SPIE, 1988, 971, 252.

**Molecular Electronic Materials - 21st Century Fine Chemicals?**

Molecular Electronic Materials
D. Bloor (University of Durham)

# Molecular Electronic Materials

D. Bloor

APPLIED PHYSICS GROUP, THE SCHOOL OF ENGINEERING AND APPLIED
SCIENCE, UNIVERSITY OF DURHAM, DURHAM DH1 3LE, UK

## 1 INTRODUCTION

The use of molecular materials for electronics and optoelectronics, both current and future, has been amply illustrated by many of the preceding articles. It is not surprising, therefore, that molecular electronics has emerged as an important area of research and application. In this process two main streams of activity have developed. It is the first of these, the use of the unique combinations of macroscopic properties of bulk molecular materials that has featured strongly in this meeting. This reflects the fact that molecular materials are already used commercially. Liquid crystals are taking an increasing fraction of the market for displays. The development of new materials and a better understanding of the basic science of liquid crystals, as described by Funada, Sage and Raynes in this volume, offers excellent prospects for further near and mid-term utilisation. Organic non-linear optical materials have been discussed in the contributions of Gray, DeMartino and Broussoux. While this work is still confined to the research laboratory the prospects for molecular materials, in particular polymers, making a breakthrough into application, in an area that has been the preserve of inorganic solids, appear to be good. The reproduction of high quality computer and video displays and optical data storage are areas where molecular materials can contribute (Hann, Vollmann this volume). This is also true of the preparation of semiconductor materials by MOCVD (Bradley). However, in many instances the molecular material is either not used in the end application or its role is purely passive. Thus a distinction has grown up between molecular materials which play active roles in device function and those which, though important in the production of devices, play either enabling or passive roles. It is the former class of molecular materials which are the concern of molecular electronics.

The second strand of molecular electronics is the utilisation of processes occurring at or near the molecular scale as a basis for novel high density electronic devices. Though such devices have been discussed in the literature most of the systems described have been conjectural. It has been easy to postulate that the ultimate limit of device size must be the molecular scale, e.g. on the basis of extrapolations of the reduction in component sizes over several decades as shown in Figure 1, but difficult to demonstrate the required molecular properties for either individual molecules or small aggregates. This situation is changing rapidly through developments in science and engineering that can provide the tools necessary to investigate processes at the molecular scale. The first steps towards the long term goal of molecular scale electronics are now being taken through multidisciplinary

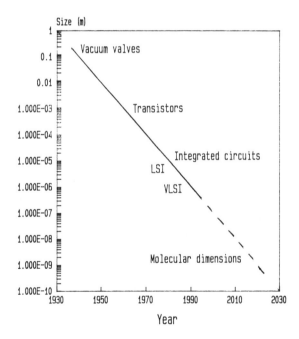

**Figure 1** Historical and extrapolated trend in the size of active electronic components.

research stimulated by initiatives in molecular and bioelectronics, e.g. in the UK and Japan.

Current research in the two strands of molecular electronics and the possible application areas are discussed below. The section on molecular materials for electronics will concentrate on materials not dealt with in detail elsewhere in this volume. That on molecular scale electronics will present possible ways towards what is an extremely long term applications goal.

## 2 MOLECULAR MATERIALS FOR ELECTRONICS

The main categories of molecular materials either utilised or of potential use in electronics are listed in Table 1.

Table 1

## MOLECULAR MATERIALS FOR ELECTRONICS

1. Organic metals and semiconductors

2. Non-linear optical materials

3. Liquid crystalline materials

4. Photo/electro-chromic materials

5. Piezo/pyro-electric materials

Of these the second and third classes have been covered by other papers in this volume as noted in the introduction. Those not dealt with in detail are the remaining three classes; organic metals and semiconductors, photo- and electro-chromic materials and piezo- and pyro-electric materials. These will be considered briefly below.

### Organic metals and semiconductors

Though much of the recent research interest has centred on highly conductive charge transfer salts and 'doped' conjugated polymers the major current application of organic semiconductors is in xerography. This requires semiconducting, or semi-insulating, materials with a strong photoconductive response to illumination. In the following an arbitrary division of the area is therefore made into (a) photo-conductors, including low molecular weight semiconductors, (b) charge-transfer salts and (c) doped conjugated polymers.

Photoconductors. Most organic compounds and polymers that contain conjugated bonds as either cyclic or linear sequences show semiconducting behaviour, i.e. conductivities in the range $10^{-7}$ to $10^2$ S/cm, which are thermally activated and enhanced when illuminated with light of an appropriate wavelength. The science and applications of these materials have been extensively reviewed[1-4]. Much effort has been expended on understanding conduction in aromatic solids[1,2]. It has been shown that trace impurities can have a profound effect on both optical and electronic properties. Thus, ultrapure materials display, at low temperature, large values of carrier mobility characteristic of true band-like motion[5]. However, at room temperature the thermally induced lattice vibrations result in strong scattering and mobilities of 1 cm$^2$ V$^{-1}$s$^{-1}$ or less. In general photocarrier generation efficiencies are low since coulomb interaction of photoexcited carriers of opposite sign leads to rapid geminate recombination, which dominates over charge separation. These properties are markedly inferior to those of inorganic semiconductors.

Phthalocyanines have been thoroughly studied but have similar properties and in addition are sensitive to atmospheric gases[3]. Despite these problems polymers such as polyvinylcarbazole (PVK) have found use in xerography. Images are formed by photoconductive discharge of a corona charged surface. To achieve this materials are required to have a low dark conductivity but a high quantum efficiency for carrier generation and good carrier mobility. Many polymers, either in pure form or with suitable additives, have been studied in this context[4]. In general these materials exhibit hopping conductivity in which carriers hop from one local site to another, e.g. holes hopping between vinylcarbazole units in PVK. Additives can greatly enhance photogeneration efficiency and carrier mobility, values of $10^{-6}$ - $10^{-4}$ cm$^2$ V$^{-1}$ s$^{-1}$ are adequate though higher values are desirable. Photogeneration is often aided by photoinjection across an interface, which provides a barrier against recombination.

Recently promising device characteristics have been achieved in all organic photovoltaic cells and light emitting diodes[6-8]. These developments derive from work in the area of xerography, i.e. utilise materials with a high photocarrier injection efficiency. However, though the devices obtained were comparable in performance with the earliest semiconductor analogues they fall well short of current semiconductor technology. It, therefore, is unlikely that they will impact technology in the short term.

Charge-transfer salts. These first attracted attention over two decades ago as purely organic materials with metallic levels of electrical conductivity[9]. Many examples of this class of materials have been studied and conductivities in good single crystal samples can reach several thousand Siemens/cm. However, the production of high quality crystals is often difficult and their use restricted. They have been incorporated in capacitors as a protective layer which is destroyed in the event of a short circuit thus isolating the area of breakdown.

Dispersions of fine CT salt crystals produced during the casting from a common solvent, reticulate doping, results in conductive composites[10]. Only small fractions of CT salt are required to give a conducting material so that the intrinsic properties of strength and flexibility of the polymer are not adversely affected. This avoids problems encountered with polymers loaded with either carbon or metallic particles where much higher loadings are required. Transparent antistatic polymer sheet has been produced commercially with TTF:TCNQ reticulate doping[11]. In this instance the critical factor was the development of synthetic routes which enabled the components of the CT salt to be produced at low cost.

Doped Conjugated Polymers. The potential use of highly conductive polymers have excited interest since the first discovery of the enhancement of the conductivity of polyacetylene on exposure to iodine. End uses in batteries, radiofrequency screening, solar cells, etc. have been proposed. It has, however, taken more than a decade to progress from over-optimistic projections to a few emerging applications. There are obvious reasons for this. First most of the materials studied were intractable, particularly when conductive, and unstable when exposed to air. Polyacetylene rapidly oxidises, a fact which invalidates much of the early literature on this material since little care was taken with sample handling procedures.

Progress has been made in (a) achieving higher levels of conductivity, (b) greater stability in a normal ambient and (c) producing melt and solution processible materials. The highest conductivities reported for polyacetylene,

$>10^5$ S/cm, are similar to those of the best metals. Polymers such as polypyrrole and polyaniline have been obtained with projected lifetimes of more than 10 years in air. Polymers modified with alkyl pendant groups can be both melt and solution processed. Thus significant advances have been made. However, no single polymer yet shows optimum behaviour. Polyacetylene remains intractable when doped and unstable in air. Processable polymers with better stability show much lower conductivity, typically of the order $10^2$ S/cm.

Despite this these materials are now being used commercially and the prospects for the future appear good. Batteries incorporating polyaniline as an electrode are now on sale and batteries and capacitors, with attractive device characteristics, using polypyrrole electrodes are nearing commercialisation[12]. Better understanding and control of materials properties have aided such developments. Progress with polyacetylene provides an example of how better synthetic routes have been the key to recent advances. First the development of a route via a precursor polymer, the 'Durham' route (Figure 2), enabled purer polyacetylene with more carefully controlled morphology to be obtained[13]. Using this route thin films of precise thickness can be obtained and the properties of the semiconducting polymer explored by clean injection of charge, in contrast to chemical 'doping', in MISFET structures[14]. This has revealed a new device physics and effects that do not occur in inorganic semiconductors, i.e. modulation of infrared (mid band-gap) absorption by an applied potential. Such unique properties may lead to

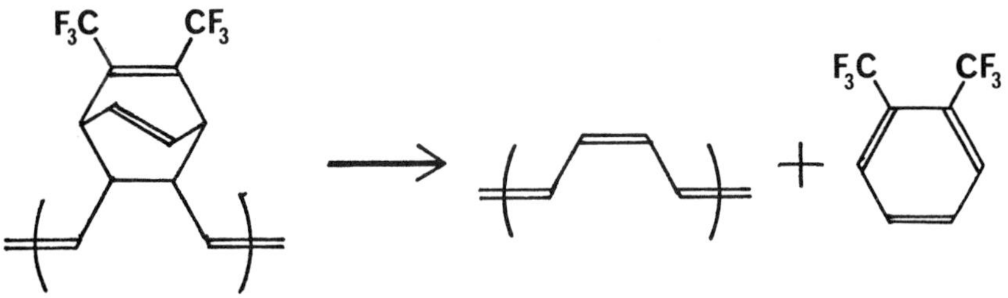

Figure 2   Precursor polymer (left) and its thermal conversion to polyacetylene.

eventual applications once the problem of handling and encapsulation are solved.

Another significant step towards applicable materials has been the development of well controlled metathesis polymerization of cyclo-octatetraene and related monomers with specific organo-metallic catalysts[15]. This provides a route to polymers with narrow molecular weight distributions, soluble polymers, etc. and should have a significant long term impact.

## Photo- and Electro-Chromic Materials

The application prospects of photo- and electro-chromic materials in displays, optical recording, etc. have also been hindered by problems of long term stability. Chemical degradation continues to be a problem in electro-chromic polymers such as polyaniline but has been largely eliminated in the case of photo-chromics by careful chemical design of the molecules[16]. This has also eliminated the thermally induced reverse reactions which adversely affect many photo-chromic materials. A range of applications have been considered ranging from optical storage through security printing to novelty clothing. Significant use in areas of electronics and opto-electronics has not yet emerged from the laboratory though non-linear optical effects and the fabrication of optical waveguides have been demonstrated.

## Piezo- and Pyro-Electric Materials

The piezo-electric properties of polyvinylidene fluoride (PVDF) are exploited in transduction of acoustic and ultrasonic signals. Because of the excellent matching between PVDF and water and living tissue it is principally employed for hydrophones and medical diagnostic equipment. It is commercially available as granules, sheet, poled sheet and poled and electroded material ready for end use. The challenge of producing polymers with properties better than PVDF has stimulated much synthetic effect but without significant success to date.

There also remains a need for good pyroelectric materials. Design considerations for pyro-electric sensors indicates an optimum thickness in a range accessible to Langmuir-Blodgett (LB) films[17]. The fabrication of polar LB films from alternate layers of different molecules has been demonstrated, initially in the context of non-linear optical materials. There are, therefore, good prospects for the development of organic pyro-electric materials capable of competing with currently utilised ceramics.

## 3 ELECTRONICS AT THE MOLECULAR LEVEL

It was recognised that with the advent of integrated electronic devices extrapolation of the reduction in size of active components would lead eventually to the molecular scale, cf. Figure 1. Early work on organic semiconductors revealed problems of material stability, purity and low carrier mobilities discussed above. However, there has been a continuing interest in the concept of the use of molecules for very small scale devices[18-22]. This has sprung from the increasing ability of chemists to synthesise complex functional molecules[22] and an interest in novel computational architectures[19,21]. There has always been the tantalising proof of principle offered by living systems presenting a challenge for scientists to emulate.

Despite this continuing interest progress has been slow. This is to some extent due to the gulf between the concept of a molecularly based computer and our ability to manipulate molecules and investigate processes occurring at the molecular scale. Thus we are faced with several important questions:

1. What materials should be studied?
2. How can individual molecules be assembled into complex supramolecular structures?
3. Can changes in the charge, energy or shape of individual molecules be detected and studied?
4. Can processes occurring at the molecular scale be controlled or at least understood sufficiently well to enable them to be used for the transfer and manipulation of information?
5. What would be the most appropriate way to implement computation using molecular scale processes?

These questions illustrate another point. The science involved in progress towards molecular scale electronics spans several disciplines and requires the establishment of multi-disciplinary research groups. Such research is stimulating because of the cross fertilisation of ideas which can occur but can be time consuming in its initial phases. Attracting established research workers into areas away from their immediate expertise and breaking down the communications barrier created by the jargon extant in each area requires considerable time and effort. However, developments in a number of disciplines are converging at the molecular scale. This together with worldwide support provided by national research programmes is providing the impetus to overcome these difficulties.

In the following sections the research that is being initiated to address the questions posed above will be discussed briefly.

## Materials for Molecular Scale Electronics

Much of the emphasis on materials for molecular scale electronics has been on systems which act as direct analogues of conventional electronic and opto-electronic switches (gates)[18,22]. This can be questioned since it is too closely connected to existing technology. Though it is now possible to micro-fabricate vacuum valves in silicon, transistors did not originally copy the structure of vacuum valves; little progress would have been achieved had research attempted to do so at that time. There are other reasons discussed below why the simple mimicking of Si technology is not necessarily a relevant approach for molecular electronics. Nature provides excellent examples and with the exception of photoreceptors utilises ions and molecules in information transport and manipulation.

Against this background two potential areas of interest are emerging as summarised in Table 2. These utilise

**Table 2**

## STRUCTURES FOR MOLECULAR SCALE ELECTRONICS

| Rigid – Solid State | Fluid – Solution |
|---|---|
| Rigid Molecules | Flexible molecules |
| Fixed arrays:<br>  Two dimensional<br>  Three dimensional | Dynamic arrays:<br>  Vesicles<br>  Membranes |
| Bi- or multi-stability:<br>  Conformation<br>  Electronic states | Bi- and multi-stability:<br>  Molecular shape<br>  Potentials |
| Transport:<br>  Electrons<br>  Excitons | Transport:<br>  Ions<br>  Molecules |

first, advances in synthetic chemistry to approach molecular scale electronics without explicit reference to natural systems and secondly, lean heavily on the examples provided by nature but attempt to abstract simpler synthetic systems which retain functionality in less complex structures.

The route involving continuous, i.e. solid state, structures is based on relatively rigid molecular units which can be formed into two dimensional, e.g. monolayer, or three dimensional arrays. Functionality will be provided by the existence of two or more distinct molecular states. While most emphasis to date has been placed on electronic states and the generation of free carriers the possibilities afforded by a change in molecular conformation cannot be ignored. These can be coupled with changes in electronic state to help stabilise two or more states of the molecule. Alternatively purely mechanical effects could be used where changes transmitted through a molecular or supramolecular framework can affect remote chromophores. Thus in addition to information transport by motion of carriers, both electrons and holes, or excitons, e.g. long lived triplet states, direct mechanical linkage could also be employed.

Using natural systems as guide lines one moves to dynamic structures and fluid environments. Analogues of natural membranes and their active components will utilise flexible molecules to construct vesicles and membranes in which the molecules are not rigidly fixed. Functionality is achieved through changes of molecular shape induced either by chemical agents or applied potentials, e.g. molecules forming ion channels. These can produce detectable changes in membrane potential. An emphasis emerges therefore on electrochemical systems and their behaviour. Information transport is achieved by either ion or molecular transport.

Assembly and Investigation

Table 3 lists some of the tools that will be important in the manipulation of molecules either individually or in controlled aggregates. Of prime importance has been the development of microscopic techniques with atomic scale resolution capable of imaging individual molecules[23-25]. In the scanning tunnelling microscope (STM) this is achieved through the exponential dependence of tunnelling currents on the tunnelling distance. Thus a single atomic centre on a tungsten tip will be favoured giving atomic scale resolution. This method allows local injection of charge

*Molecular Electronic Materials* 275

**Table 3**

which has been used through the measurement of resulting emission for molecular scale analysis. The injection of electrons may damage organics but the atomic force microscope (AFM) which relies on the repulsive potential at close distances avoids this problem. Such 'damage' has, however, been used to trigger reactions in single molecules. These techniques are still new and much remains to be learnt about how to probe molecules and obtain unambiguous results. The related technique of near field optical microscopy (NFOM) has lower resolution but allows the interaction of photons with molecular aggregates to be studied.

Techniques of sub-micron lithography are now capable of producing small scale structures which can be used either as microelectrodes or as sites for molecular adsorption and binding or both. Thus it provides an alternative route to the study of small arrays of molecules.

There are already routes available to more extended structures. Surface adsorption can provide regular arrays of large numbers of molecules. The LB film technique has been used to make alternating layer structures with specific functionalities. LB films can be used to make bilayer containing biologically derived materials and their analogues. In principle more complex multi-monolayer structures can be produced. Finally controlled reaction at surfaces can also provide both two and three dimensional molecular structures comprising regions with different constituents and functionality.

Control and Computation

These have been areas of speculation and will remain so until the advances described above lead to positive demonstration of molecular functions that can be used for electronic purposes. The problems being posed by reducing component sizes in silicon can provide useful lessons. As device density increases there is a need to simplify interconnections to keep overall complexity within manageable bounds. Integrated circuits though performing complex functions have relatively few inputs and outputs. Thus, once device physics is understood system behaviour can be determined and direct intervention is not needed during information handling. By analogy, though direct measurement of molecular systems is needed now, once an adequate understanding of molecular scale properties is established device and system architectures can be designed within minimal input and output.

Thus though much basic research is required before realistic system concepts can be developed there are some general constraints which are now apparent. The considerations mentioned above together with the possibility of producing molecular arrays suggests that devices based on cellular automaton models are a possibility. Large areas of mono- or multi-layers could also derive power and timing information from laser pulse trains. Thus one can visualise how rigid-synthetic structures might be used to produce molecular electronic systems. The use of preformed structures would avoid problems inherent in the molecular switch approach, in particular, how specific structures of known functionality could be formed for very large numbers of such molecules.

If systems using either ion or molecular transport are to be used then there must be recourse to either parallel or multiply connected computational architectures to compensate for the inherent slow speed of such processes. How far parallelism can compensate remains an open question. Some advantage may be gained from multi-micro-electrode arrays fabricated by sub-micron lithography. In fluid systems chemical reactions can provide energy input and to some extent timing. Thus it seems likely that the structures and architecture of devices derived from biological analogues will be very different from any solid state counterparts.

4 CONCLUSIONS

Molecular materials have already found applications in electronics. In such developments the refinement of properties to achieve improved performance in particular end uses present a challenge to the synthetic chemist. Their purification to the levels essential in most applications is also far from trivial. However, as unusual properties unique to molecular materials are discovered there will be continuing interest and utilisation.

The prospects for molecular scale electronics are more difficult to define. Research is still in its early stages and applications seem unlikely, in the absence of unforseen discoveries, until well into the next century. However the convergence of different disciplines, as illustrated in Figure 3, has created an opportunity to explore experimentally properties of materials at the molecular scale. In particular the development of atomic resolution microscopy and its use to excite and modify molecules provides a tool without which progress would be extremely difficult. The interdisciplinary research, which plays an important role in underpining the use of molecular materials in current technology, is even more vital for studies at the molecular scale. Such comprehensive team

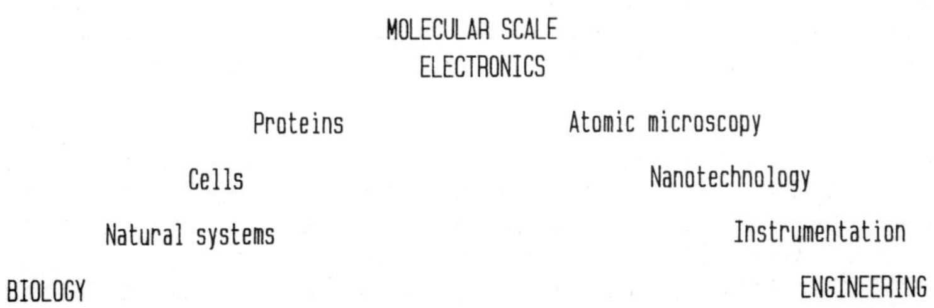

| CHEMISTRY | PHYSICS |
|---|---|
| Atoms | Solids |
| Molecules | Semiconductors |
| Supramolecular structures | Low dimensional structures |
| MOLECULAR SCALE ELECTRONICS | |
| Proteins | Atomic microscopy |
| Cells | Nanotechnology |
| Natural systems | Instrumentation |
| BIOLOGY | ENGINEERING |

**Figure 3** Molecular scale electronics as the point of convergence of developments originating in different disciplines

efforts are now being established and will inevitably produce good science and further our understanding. Only time will tell what the impact on technology will be.

REFERENCES

1. M. E. Pope and C. E. Swenberg, 'Electronic Processes in Organic Crystals', Oxford University Press, Oxford, 1982.
2. E. A. Silinsh, 'Organic Molecular Crystals', Springer Verlag, Berlin, 1980.
3. J. Simon and J-J. Andre, 'Molecular Semiconductors', Springer Verlag, Berlin, 1985.
4. H. W. Gibson, Polymer, 1984, 25, 3.
5. W. Warta and N. Karl, Phys. Rev. B, 1985, 32, 1172.
6. C. W. Tang, Appl. Phys. Lett., 1986, 48, 183.
7. C. W. Tang, S. A. VanSlyke and C. H. Chen, J. Appl. Phys., 1989, 65, 3610.
8. F. F. So and S. R. Forrest, Appl. Phys. Lett., 1988, 52, 1341.
9. J. B. Torrence, Acc. Chem. Res., 1979, 12, 79.
10. A. Tracz, J. K. Jeszka, E. E. Shafee, J. Olanski and M. Kryszewski, J. Phys. D., 1986, 19, 1047.
11. J. Hocker, F. Jonas and H. K. Muller, Angew. Macromol. Chemie, 1986, 145/146, 191.
12. D. Naegele, Springer Ser. in Sol.State SCi, 1989, 91, 428.
13. J. H. Edwards, J. W. Feast and D. C. Bott, Polymer, 1984, 25, 395.
14. J. H. Burroughes, C. A. Jones and R. H. Friend, Nature, 1988, 335, 137.
15. R. Grubbs in 'Conjugated Polymeric Materials', Ed. J. L. Bredas, NATO ASI Series in press.
16. H. G. Heller, in 'Fine Chemicals for the Electronics Industry', Ed. P. Bamfield, Royal Society of Chem., London, 1986, p.120.
17. C. A. Jones, M. C. Petty and G. G. Roberts, IEE Trans. Ultrasonics, 1988, 35, 736.
18. F. L. Carter, Ed., 'Molecular Electronic Devices', Marcel Dekker, New York, 1982.
19. M. Conrad, Commun. A.C.M., 1985, 28, 464.
20. M. La Brecque, Mosaic, 1989, 20, 16 and 28.
21. J. R. Barker, Hybrid Circuits, 1987, No14, 19.
22. J. J. Hopfield, J. N. Onuchic and D. N. Beratan, Science, 1988, 241, 817.
23. J. S. Foster, J. E. Frommer and P. C. Arnett, Nature, 1988, 331, 324.
24. P. K. Hansma, V. B. Elings, O. Marti and C. E. Bracker, Science, 1988, 242, 209.
25. M. Amrein, R. Durr, A. Stasiak, H. Gross and G. Travaglini, Science, 1989, 243, 1708.

# Subject Index

Aggregates
 molecular,
  266, 274, 275
Alloys, 36, 37
 rare earth/transition metal,
  198, 199, 200
 solid solutions, 37
Aluminium indium arsenide, 50
Aluminium nitride, 52
Amines
 complex, 18
AM-LCD (see display, liquid
  crystal, active matrix)
Antiferroelectric
 phase, 107, 109
Arsenic, 36, 43, 44, 45
Arsine, 12, 25, 49, 51, 52
Azides
 aromatic, 18, 19

Babinet compensator, 108
Baiyunebo ore process, 73, 75
Bastnasite, 68
 processing, 72
Bioelectronics, 267

Cadmium mercury telluride, 38, 50
Cadmium selenium telluride, 50, 53
Cadmium selenide, 38, 53, 136
Cadmium sulphide, 53
Cadmium telluride, 38
Capacitors, 269, 270
Casting resin, 163-182

Casting resin
 (continued)
 flexible, 163
 polyurethane insulators, 163
 room temperature curing, 163
 tough, 163, 168
Cathode ray tube, 100, 101, 102, 111, 114, 130
 flat, 100, 101, 102
 invention, 97
CD-ROM (see compact disc - read only memory)
Charge-transfer salts, 267, 269
 TTF:TCNQ, 269
Chemical vapour deposition (CVD), 7, 8, 15, 27, 29, 31
 metallo organic, 49-59
 photoenhanced, 30, 32
 plasma, 30, 32
Chip
 integrated, 3, 4, 5
CMY, 147, 149
Colour
 primary additive, 147
 primary subtractive, 147, 158
Compact disc
 audio, 183, 184
 video, 183
 recording processors
 liquid crystal phase switching, 184
 metal film alloying, 184
 photochromic colour change, 184, 185

Compact disc-read only
 memory, 183, 184
Composites, conducting, 269
Computer
 personal
 compatibility with display,
 103
 display market trends, 103
 world market, 101
Conduction band, 36, 39
Continuous tone image, 148
Copolymer
 liquid crystal
 lateral/terminal side chain,
 206
CRT (see cathode ray tube)
Cyanobiphenyls
 development, 105
 properties, 105,
Cyclo-octatetraene,
 metathesis polymerisation,
 271
Cyclodextrin
 in electrochromic displays,
 115

D2T2 (see dye diffusion
 thermal transfer printing)
DCM-LCD (see display, liquid
 crystal, dynamic
 scattering mode)
Deposition, 7, 9, 11-16, 25, 26
 tungsten, 31
Detector
 infra red, 38, 50
Device
 biological analogues, 277
 imaging, 50
 input, 97, 98
 laser addressed storage, 214
 memory, 97, 98
 molecular electronic, 266
 molecular materials, 265, 266

Device (continued)
 output, 97, 98
 processor, 97, 98
 semiconductor, 3-5, 15, 24,
 25, 28, 32
 thermally addressed storage,
 214
 smectic A storage, 213
 transmission, 97, 98
Diborane, 12, 26, 27, 29
Dielectric, 14
Diels-Alder coupling
 use in liquid crystal
 synthesis, 123
Differential scanning
 calorimetry
 use in polymer liquid crystal
 research, 208, 209
Diisocyanatodiphenyl methane,
 164
 polyetherpolyol resin, 164,
 168
 prepolymers, 165
Direct random access memory,
 99, 100
 IG, 99, 100
 16-64M, 99, 100
Display
 compatibility with personal
 computer, 103
Display, elastomer, 100
Display, electret, 100
Display, electrochromic, 100,
 101, 102, 115, 121
 applications, 116
 lifetime, 115, 116
 limitations, 116
 RGB, 116
Display, electroluminescent,
 50, 100
Display, electrophoretic, 100,
 101, 116
Display, electroscopic, 100

Display, high information
  content, 101, 102, 104,
  107, 135
  materials, 105
Display, liquid crystal, 100,
  101, 102, 116-129, 130
  active matrix, 105, 108, 109,
    117, 124, 128, 135
  addressing, 127
  characteristics, 110
  cost, 110
  full colour, 110
  liquid crystal material
    properties, 105
  use of fluorine substituted
    liquid crystals in, 127, 128
  construction, 133, 134
  double layer supertwisted
    nematic, 108, 137, 138
  characteristics, 97
  full colour, 109
  structure, 108, 109
  dynamic scattering mode, 104,
    117
  electroclinic, 117
  ferroelectric, 105, 107, 109,
    116, 124, 138, 140
  alignment, 140, 141, 143
  biaxiality, 143, 144, 145
  characteristics, 109
  molecular arrangement, 125,
    141, 142
  operation, 140
  physics, 141, 144
  response time, 123, 145
  spontaneous polarisation,
    144
  twisted director profile,
    143
  X-ray studies, 141
  film compensated
    supertwisted nematic, 109
  information level, 120

Display, liquid crystal
  (continued)
  investment in Japan, 101
  materials, 104
  multiplexed, 117
  performance, 115
  pixel content, 104
  polymer dispersed, 117
  suitability for HDTV, 104
  supertwisted nematic, 105,
    108, 109, 120, 121, 136-
    138
  adhesive, 120
  appearance and quality 120,
    121
  characteristics, 108
  contrast, 121
  full colour, 138
  glass, 120
  grey scale, 138
  liquid crystal properties,
    120, 121
  multiplexing scheme, 122
  optical properties, 136
  polariser, 120
  polyimide, 120
  retardation film, 109, 137,
    138
  speed, 138
  transmission spectrum, 137
  voltage dependence, 137
  thermally addressed smectic,
    117
  thin film transistor, 110, 136
    cost, 110
  tunable birefringence, 117
  twisted nematic, 104, 117,
    121, 124, 126
  driving voltage, 124, 126
  limitations, 135
  liquid crystal properties,
    121
  operation, 134, 135

Display, liquid crystal
  (continued)
  twisted nematic (continued)
    optics, 135
    voltage ratio, 120
    x-y matrix, 107, 108
Display, magneto-optic, 100
Display, micromechanical, 116
Display, passive, 114, 115
Display, plasma, 100, 101, 102
Display
  principles, 100
  properties, 114
  requirements, 99
Display, suspended particle, 100
Display, vacuum fluorescent, 100, 101, 102
DRAM (see direct random access memory)
DSTN-LCD (see display, liquid crystal, double layer supertwisted nematic)
Dye
  infrared absorbing, 183
  NIR
    dithiolatonickel complexes, 196, 197, 200
    metal complexes, 196, 197
    methine derivatives, 188-191, 200
    miscellaneous, 197
    naphthalocyanine derivatives, 192, 193, 194, 200
    1,4-naphthoquinones, 195
    phenothiazines, 196
    phthalocyanine derivatives, 191-192, 196, 200
    quinoid polynuclear aromatic, 194
    selenazine quinones, 196
    NIR sensitive, 188-197, 199

Dye (continued)
  NIR (continued)
  Dye diffusion thermal transfer printing, 147-159
    colour ribbon, 148, 149, 151, 152, 153
    diffusion coefficient, 154, 156
    digitisation, 150
    dye, 152, 153
    dye layer, 151
    dye performance characteristics, 152
    heat generation, 148
    material structure, 151
    memory, 150
    modelling, 153-158
    partition coefficient, 153, 156
    receiver paper, 148, 150, 151, 152, 153
    reflection density, 157
    substrates, 152
    thermal head, 148, 149, 151, 155
    thermal head-temperature distribution, 155, 156
    thermal modelling, 155
    transmission density, 155

E-O (see electro-optic)
ECD (see display, electrochromic)
EDRAW (see erasable direct read after write)
EL (see display, electroluminescent)
Elastic constant, 117, 132
  splay, 132
  twist, 132
  bend, 132
Electrochromics, 267, 271
Electronics, molecular

## Subject Index

Electronics, molecular
(continued)
  molecular materials for,
  265-272, 277
  molecular level/molecular
  scale, 266, 272-277
Electron beam
  discovery, 97
Electron mobility, 50, 51
Electro-optic
  constant, 231, 232, 234,
  235, 251
  device figure of merit
  (see electro-optic
  constant)
  devices, 226, 233
  modulation, 247, 251-254,
  256-258
Elements
  groups II, III, V, VI
  arrangement, 37
Energy band gap, 36, 37, 38, 39
EPD (see display,
  electrophoretic)
EPIC (see reaction resin, epoxy
  with isocyanate based)
Epitaxy
  processes, 51, 54
Erasable direct read after write
  183, 184, 185, 199, 200
  physical marking process, 185
ESD (see display, electroscopic)
Eye
  colour sensor, 147

Faraday effect
  magneto-optical, 198
FE-LCD (see display,
  liquid crystal, ferroelectric)
Ferrites, 199
FET (see field effect
  transistor)
Field effect transistor, 50, 136

Fluorine
  atomic, 23
F-STN-LCD (see display, liquid
  crystal, film compensated
  supertwisted nematic)

Gallium, 49
Gallium aluminium arsenide,
  38, 50
Gallium arsenide,
  38, 49, 50, 52, 53
Gallium indium arsenide, 50
Gallium indium arsenide
  phosphide, 50
Germanium, 36, 38
Giant Electronics Technology
  Corporation, 110
Glass fibres
  fluoride, 56, 57, 92

Hall measurement, 47, 48
HDTV (see high definition
  television)
High definition television,
  101, 104, 110, 111
  applications, 104
  display requirements, 104
  industrial importance, 104
  market size, 104
High voltage arc resistance, 177
High voltage diffusion dielectric
  strength, 177
High voltage tracking resistance,
  177
Homopolymer, liquid crystal
  side chain, 206
Hull University Liquid Crystal
  Group, 123

IEC Standard
  in casting resin technology,
  168
Indium, 36, 43, 44

Indium antimonide, 38
Indium phosphide, 38, 43, 50, 51, 52
Inhibitor, 18
Insulator, 53
  glass, 176
  outdoor, 176, 178
    long term tests, 176
    short term tests, 176, 177
  porcelain, 176
  erosion
    epoxy, 178, 179
    filler, 177, 179, 180
    pitting, 177, 178, 179, 181
    plain, 178
  surface, 176
Integrated circuit
  evolution, 100
  invention, 97

JOERS (see Joint Opto-Electronics Research Scheme)
Joint Opto-Electronics Research Scheme, 52
Jones Matrix, 135, 136, 138, 143

Kerr effect
  magneto-optical, 198
Kevlar, 205

Langmuir-Blodgett (LB) films, 271, 275
Lanthanides
  heavy, 63, 66, 67, 71
  light, 63, 66, 67, 71
Laser diode, 50
Laser erasure, 183, 198
Lasers
  dye, YAG, use in second harmonic generation studies, 230, 231

Lasers (continued)
  frequency doubling, 200
  gas
    non-directly modulable, 188
    tunable, 224
Laser writing, 183, 198
LCD (see display, liquid crystal)
Layer structures, in molecular electronics
  bilayers, 275
  monolayers, 274, 276
  multilayers, 276
Lead lanthanum zirconate titanate, 85
Lead selenide, 50
Lead sulphide, 50
Lead telluride, 50
LED (see light emitting diode)
Light emitting diode, 38, 50, 100, 101, 268
Light filter
  fixed wavelength, 211
Light reflector,
  fixed wavelength, 211
Light valve, 110
LIMA measurement, 46, 47
Lithium niobate
  devices, 253, 254, 259
Liquid crystal
  alignment, 130, 132, 133
    homeotropic, 132
    homogeneous, 132
    in electric field, 130
  orientational order, 131
  amphiphilic systems, 204
  birefringence, 118
  calamitic systems, 203
  cholesteric, 131
  clearing point, 118, 119
  cyano compounds, 105, 124, 125, 126
  design rule, 117, 120
  dielectric anisotropy,

Liquid crystal (continued)
  dielectric anisotropy
    (continued)
    118, 126, 127
  dimeric arrangement, 126, 127
  dipole moment, 126
  discotic systems, 204
  elastic constants, 117, 132
  electro-optic effects, 133
  esters
    laterally fluorinated, 125
  features, 105, 106
  ferroelectric, 138, 139, 145
  chiral dopants, 126
  difluoroterphenyls, 139, 144
  origin of ferroelectricity, 138
  properties, 125, 126
  fluorine substituted, 127
  fluorine substituted in active
    matrix displays, 127, 128
  lateral fluorination, 122, 123, 124
  lyotropic, 204, 205
  nematic
    director, 132
    elastic constants, 132
    electric field effects, 144
    electric permittivity, 132, 144
    orientational elasticity, 132
    refractive indices, 132
  optimisation, 117
  phase structure, 130, 131
  physical properties, 130, 132
  physics, 130-146
  production, 122, 123, 124
  properties, 120, 121, 130, 132
  properties in displays, 120, 121
  smectic, 131
  stability, 117
  structures, 118, 119, 122
  synthesis, 123
  transition temperatures, 118, 120

Liquid crystal (continued)
  viscosity, 121
  Liquid crystal copolymer (see
    copolymer, liquid crystal)
  Liquid crystal homopolymer (see
    homopolymer, liquid crystal)
  Liquid crystal polymer (see
    polymer, liquid crystal)
Low-dimensional structures, 277
Low temperature oxide (LTO), 14, 26

Magneto-optical recording, 198, 200
Magneto-optical recording
  medium, 198
Malonate/diol polyesters, 210
Mauguin optical effect, 134
MDI (see diisocyanatodiphenyl-methane)
Membranes, 273, 274
Metalloids, 36
Metals, 36
  transition, 36
MBE (see molecular beam epitaxy)
Melinex, 152
Metal-insulator-metal, 135
Metallo-organic compounds
  precursors, 49, 50, 52, 53, 54
  purity, 50
Metal organic semiconductor field
  effect transistor (MOSFET), 4
  processing, 6
Metal oxides
  uses, 54, 55
Metallisation, 30, 31
Microcircuits, 16
Microelectronics, 3, 5, 7, 13
Microscopic techniques,
  use in molecular electronics
  atomic force
    microscope (AFM), 275

Microscopic techniques,
  use in molecular electronics
  (continued)
  near field optical microscope
    (NFOM), 275
  scanning tunnelling microscope
    (STM), 274, 275
  MIM (see metal-insulator-metal)
Ministry of International Trade
  and Industry, 110
Mobility
  carriers, 268, 272, 274
  photocarriers, 268
MOCVD (see chemical vapour
  decomposition, metallo-
  organic)
Modulators
  high speed, 224
  lithium niobate, 234, 235, 253,
    254
  Mach-Zehnder electro-optic,
    223, 247, 251-254, 256-
    258
  polymer, 234, 235, 247
Molycorp process, 73, 74, 78
Molecular beam epitaxy, 49
  metallo-organic, 52
Molecular computer, 272
MOMBE (see molecular beam
  epitaxy, metallo-organic)
Monazite, 68, 70
  processing, 76
  radioactivity, 70
Moore's law, 99

Nanotechnology, 277
NLO (see nonlinear optics)
Nonlinear optics, organic
  applications, 224, 225, 227
  β-value (see also nonlinear term)
    224, 225, 229, 237, 247
  basic concepts, 224
  copolymers, 227-229, 239,

Nonlinear optics, organic
  (continued)
  copolymers (continued)
    240, 243-247, 256, 258,
    259
  devices, 223-226, 233-235,
    237, 254-258
  donor-acceptor concept,
    237, 238
  hyperpolarisability, 237, 238,
    240, 241
  Langmuir-Blodgett films, 238
  non-centrosymmetric structure,
    225, 238, 250
  nonlinear term, 224
  polymer films
    poling, 226, 229-232, 235,
      239-241, 243-244, 256
    waveguides, 238, 252, 256,
      258
  polymers
    advantages, 225-226
    amorphous, covalently
      functionalised, 243-247
    doped, 241-243
    cross-linked, 248-250, 254
    ferroelectric, 238-240
    frequency doubling, 237, 238,
      250, 251
    guest-host systems, 240,
      241, 245, 248
    glass transition point (Tg),
      225-228, 230, 238-240, 247
    isotropic, 229
    liquid crystalline, 225-228
      238, 239
    systems, 225-228
  single crystals, 238

Optical density, 157, 158
Optical disk
  cyclic polyolefin, 188
  mastering process, 186

Subject Index

Optical disk (continued)
  polycarbonate, 186, 187
  polymethyl methacrylate, 186, 187
  production, 185
  reduction of birefringence, 187
  substrate properties, 186, 187
  write once storage materials, 188
Optical memory, 183
Optical microscopy
  use in polymer liquid crystal research, 208, 209
Optical reading systems
  integrated, 200
  laser arrays, 200
Optical recording
  reversible phase change, 199
Opto-electronics, 50, 51
Organic metals, 267
Organoarsines, 52
Organophosphines, 52

PDP (see display, plasma)
Phenylarsine, 52
Phenyl arsonic acid, 52
Phosphine, 7, 12-15, 25-27, 29
Phosphor, 50
Phosphosilicate glasses (P glasses)
  use in semiconductors, 8, 9, 12, 14, 15
Photochromics, 267, 271
Photocathode, 50
Photoconductors, 267, 268
Photo-detector, 38
Photovoltaic, 38
Photovoltaic cell, 268
Piezoelectric materials, 267, 271
  polyvinylidene fluoride (PVDF)
  use in hydrophones, 271

PLZT (see lead lanthanum zirconate titanate)
Pockels constant (see electro-optic constant)
Polyacetylene
  conductivity, 269, 270
  doped, 269, 270
  "Durham" route, 270
Polyalkene sulphones, 210
Polyaniline
  electrode material, 270
  electrochromic, 271
Polymer, liquid crystal
  future development, 220
  hybrid, 205, 206
  main chain, 204, 205, 207, 208
  applications, 208
  phase identification, 208
  rheology, 208
  side chain, 204, 205, 207, 209
  acrylates, 219
  applications, 211
  backbones, 209, 210
  elastomers, 217
  electric field poled, 216
  ferroelectric, 215
  integrated optics, 217
  Langmuir-Blodgett film, 216
  malonate/diol polyesters, 220
  membranes, 217
  methacrylates, 219
  molecular weight control, 220
  optically nonlinear media, 216
  phase identification, 209
  polysiloxane, 218
  pyroelectric, 215
  second harmonic generation, 216
  thermochromism, 212
Polymer
  photochromic, 217
Polymers
  conductive, 269-270

Polymers (continued)
  doped, conjugated, 267-270
  processing
    melt, 270
    solution, 270
Polyphosphazenes, 210
Polypyrrole
  electrode material, 270
Polysilicon
  use in semiconductors, 10-14, 27, 110
  use in liquid crystal displays, 136
Polysiloxane, 207
Printing
  D2T2 (see dye, diffusion thermal transfer printing)
Printing
  sublimation transfer, 150
  textiles, 150
Phthalocyanines, 268
Pyroelectric materials, 267, 271
PUR (see reaction resin, polyurethane based)

Rare earths, 63-94
  abundance, 65-66
  applications, 80
    electrical, 83, 88
    electronic, 83, 88
    lasers, 80, 84
    magneto-optic, 83-86
    optical, 83-84
    television, 80, 83
  ceramics, 85
  electroluminescence, 91
  electronic configurations, 64
  fluoride glass fibres, 92
  future applications, 90
  chemical analysis, 80, 81
  ion exchange, 78, 79
  magneto-optical recording,

Rare earths (continued)
  magneto-optical recording (continued)
    198, 199, 200
  minerals, 67, 71
  ores, 67
  purity, 80, 81
  rechargable batteries, 91
  reserves, 66
  solvent extraction, 78
  superconductivity, 90
  valence changes, 65
  X-ray phosphons, 84
Reaction resin, 163-182
  cycloaliphatic epoxy, 176
  high voltage diffusion dielectric strength, 181
Reaction resin, epoxy with isocyanate based
  applications, 173, 175
  catalysts, 173
  comparsion with thermosetting plastics, 170, 171
  dimensional heat stability, 170
  glass transition temperature, 170
  idealised structure, 170
  permanent service temperature, 170, 173
  pre-reaction storage stability, 172
  ready to use formulations, 172
  storage stability, 172
  thermal properties, 174
  properties
    dielectric strength, 175
    electrical, 173
    flame retardance, 170, 175
    flexural strength, 174, 175
    heat class, 173
    mechanical, 173
    tensile strength, 174
    thermal, 170, 174

## Subject Index

Reaction resin (continued)
  properties (continued)
    thermal stability, 174
  resin
    casting, 172
    dipping, 172, 173
    impregnating, 172, 173
    laminating, 173
    solid, 172
    solid, solubility in organic solvents, 172
Reaction resin, polyurethane based
  aliphatic, 176
  applications, 163, 164
  approved standard systems, 164
  comparison of properties, 165, 166
  cycloaliphatic
    applications, 176
    erosion, 180, 181
    high voltage arc resistance, 177
    high voltage diffusion dielectric strength, 180
    high voltage tracking resistance, 177
    use in outdoor insulators, 176
    use in outdoor transformers, 176
Recording process
  reflectivity, 184, 196
Resists, 16-19, 28
  polymers, 16, 17
  cyclised isoprene, 19
  novolac, 18
RETM (see alloy, rare earth, transition metal)
Reversible data storage, 197
  magneto-optical effect, 197
RGB, 109, 116, 147
Rhone-Poulenc process, 76-78

SHG (see second harmonic generation)
Second harmonic generation, 230, 231, 239, 240, 242, 245-250, 255-256
Second order activity (see second order property)
Second order property ($X^{(2)}$) 224, 225, 229-231, 233, 237, 238, 250, 251
Second order susceptibility (see second order property)
Semiconductor, 36, 38, 39, 40
  analytical techniques, 43
  "atomic engineering" techniques, 44, 45
  bulk crystals, 44
  compound, 40, 44, 45, 49, 50, 52, 53
  doping, 5, 11-15, 26-27
  effect of particulate contamination on purity, 41
  electronic grade process chemicals, 5, 6, 31
  epilayer fabrication, 45, 47
  epitaxy, 44, 45
  etching, 7, 19-24
  fabrication, 5, 7, 9, 16, 19, 24, 32, 45
  impurity level, 42
  inorganic, 268, 270
  molecular scale electronic, 277
  organic, 267-269, 272
  polymer, 268, 270
  photolithography, 7, 16, 19, 21, 27-29
  purity, 40-45
  single crystals, 44
    growth technique, 44
  sputtering, 7, 20-21, 30
  wafers, 44
Semiconductor lasers
  IR transmission, 199

Silanes, 7-11, 13, 16, 25-29
Silica flour
 use in casting resins, 177, 179, 180
Silicates
 use in semiconductors, 8, 13
Silicon, 36, 38, 40, 50
Silicon, amorphous, 110, 136
 use in liquid crystal displays, 136
Silicon processing
 acids, 28, 29
 safer alternatives for, 25-29, 32
 2-ethoxyethyl acetate (2-EEA), 28
 glycol ethers, 28
 propylene glycol monomethyl ether acetate, 28
 tetraethyl orthosilicate (TEDS), 26, 27, 30
 trichloroethane (TCA), 27, 29, 30
 trimethyl borate (TMB), 27, 28, 30
Silicon, use in semiconductors, 3-5, 7-9, 12, 15, 19, 23-26, 28, 29, 31, 32
Smectic A, 105
Smectic C, 105, 138
Solar cell, 50
Sol-gel process, 54
SPD (see display, suspended particle)
Spin coating, 16
Spontaneous polarisation, 107, 125, 138, 139, 144
 in cyanohydrin esters, 125
SSF-LCD (see display, liquid crystal, ferroelectric surface stabilised)
STN-LCD (see display, liquid crystal, supertwisted nematic)

Sub-micron lithography
 use of, in molecular electronics, 275, 276
Supramolecular structures, 272, 274, 277
Switches
 electronic, 273
 molecular, 276
 optical, 224, 234
 optoelectronic, 273

Terfenol, 86
Terphenyls, 124, 125
 viscosity, 125
Tetramethyl cyclotetrasiloxane (TOMCATSTM), 26, 27
TFT (see thin film transistor)
TFT-LCD (see display, liquid crystal, thin film transistor)
Thin film transistor, 110, 136
 amorphous silicon, 136
 fabrication, 110
 polysilicon, 136
Third order effect
 (see third order property)
Third order property ($X^{(3)}$), 233
Tin lead telluride, 50
Tin sulphide, 50
Tin telluride, 50
TN-LCD (see display, liquid crystal, twisted nematic)
Transistor, 3, 4, 7, 9
 invention, 97
Transition metals
 magneto-optical recording, 198, 200
Transport
 electronic, 273
 excitonic, 273
 ionic, 273, 274, 276
 information, 273, 274
 molecular, 274, 276

## Subject Index

Trialkyls of aluminium, gallium, indium, 51
volatile adducts, 51, 52
Tri-t-butyl phosphine (TBP), 26
Trimethyl gallium, 49, 50, 51, 52
Trimethyl phosphite (TMPI), 26, 27, 30
Triode tube
invention, 97

Urethane structure
formulation, 164, 165

Valence band, 36, 39
electrons, 37
VDE standard
in casting resin technology, 176, 178
Very large scale integration, 97, 101
Vesicles, 273, 274
VFD (see display, vacuum fluorescent)

Video camera
colour sensor, 147
Viologen, 115
VLSI (see very large scale integration)
Waveguide, 50, 56
Wittig coupling
use in liquid crystal synthesis, 123
WORM (see write once read many)
Write once read many, 183, 184, 192, 193, 194, 196, 199

Xenotime, 71
processing, 76
Xerography, 267, 268
X-ray diffraction,
use in liquid crystal polymer research, 208, 209
Zinc cadmium sulphide, 38
Zinc selenide, 38, 53
Zinc sulphide, 38
Zinc sulphur selenide, 50